SERIES ON

Water: Emerging Issues and Innovative Responses

PLASTIC POLLUTION

Nature Based Solutions and Effective Governance

Editors

Gail Krantzberg
McMaster University
Hamilton, ON
Canada

Savitri Jetoo
Adjunct Professor
Abo Akademi University
Turku, Finland

Velma I. Grover
McMaster University
Hamilton, ON
Canada

Sandhya Babel
Professor
School of Bio-Chemical Engineering and Technology (BCET)
Sirindhorn International Institute of Technology (SIIT)
Pathum Thani, Thailand

CRC Press is an imprint of the
Taylor & Francis Group, an **informa** business
A SCIENCE PUBLISHERS BOOK

Cover credit:
Cover illustrations reproduced by kind courtesy of Dr. Gail Krantzberg and Dr. Sandhya Babel.

First edition published 2023
by CRC Press
6000 Broken Sound Parkway NW, Suite 300, Boca Raton, FL 33487-2742

and by CRC Press
4 Park Square, Milton Park, Abingdon, Oxon, OX14 4RN

CRC Press is an imprint of Taylor & Francis Group, LLC

Library of Congress Cataloging-in-Publication Data (applied for)

ISBN: 978-0-367-68480-8 (hbk)
ISBN: 978-0-367-68481-5 (pbk)
ISBN: 978-1-003-13774-0 (ebk)

DOI: 10.1201/9781003137740

Typeset in Times New Roman
by Radiant Productions

Preface to the Series

Water is the lifeline for all life on the planet and is intricately linked with food security, energy security and sustainable development. Water is the core of sustainable development and thus maintaining water quality is important for maintaining life on the planet. To support our ongoing development and provide comforts, we have developed many chemical substances that are harmful to the environment and one such product is plastic. The focus of this "water series" is to explore innovative solutions of treating water pollution, better governance structure and other institutional framework, good legal structure, in addition to increased investment in water sector.

Plastic products are widely used in our daily life due to their unique properties, ease, and convenience. When we look around, their presence is felt almost everywhere in our life such as plastic bottles (less risk for breaking and are lightweight), plastic bags, toys, kitchen utensils (Styrofoam cups, trays, food storage and packaging that preserves the shelf life of food), fishing nets, micro plastics in cosmetics, toothpaste and now disposable masks from the Covid-19 pandemic. As such, plastic has become one of the most pressing environmental threats of our time. The focus of this book in the water series is to explore nature based solutions and effective governance for anthropogenic plastic pollution.

Velma I. Grover

Preface

Plastics are lightweight polymers made from petro-chemical base. It is used in most of the things we use starting from toothpaste, clothes, makeup, computers, cars, etc., and is used is most of the things we do: building and construction, agriculture, healthcare, food distribution, transport, textile, electronics, communications and the list goes on. Unfortunately, it takes decades to breakdown and most plastic products do not fully degrade. With nearly a century of use of plastics, they have been detected in the air we breathe and water we drink (even in the deepest parts of ocean and Arctic Sea ice).

Generally when one hears the word plastic waste, the image of plastic bottles come to mind but when one talks about plastic waste (and microplastic waste) the list is longer as it includes synthetic fibres, disposable cups, plates, cutlery, straws, cigarette butts, plastic bags, packaging, protective mesh, and disposable health care instruments such as syringes and gloves. Although plastics have become foundational in our lives and have revolutionized the way live, we are continually vulnerable to plastics due to many shortfalls in the recycling process. The focus of this book is not just to highlight the problem of plastics, its definition, and how plastic pollution is impacting human health and environment but also to look at some best practices in both natural solutions and in the field of law and policies. The first section of the book focuses on plastic pollution—it's origin, relationship to climate change, linear/circular economy, followed by sustainable plastics, scientific solutions, and how policies can address plastic pollution. The main message in the end is that plastic waste is a complex problem and needs to be addressed more holistically. This includes looking at better designs, more sustainable feedstocks, and partnerships between various stakeholders, including the production industry, packaging industry, users, academicians/researchers, policymakers, and waste management industry. The principle is to have improved plastic design and products to facilitate reuse, repair, and recycling as suggested in the "European Strategy for Plastics in Circular Economy".

Velma I. Grover
Gail Krantzberg
Savitri Jetoo
Sandhya Babel

Contents

Chapter 1

Plastics

Nature-Based Solutions and Effective Governance

Velma I. Grover,[1,*] *Gail Krantzberg,*[1] *Savitri Jetoo*[2]
and *Sandhya Babel*[3]

INTRODUCTION

Plastic products are widely used in our daily life due to their unique properties, ease, and convenience. When we look around, their presence is felt almost everywhere in our life such as plastic bottles (less risk for breaking and are lightweight), plastic bags, toys, kitchen utensils (Styrofoam cups, trays, food storage and packaging that preserves the shelf life of food), fishing nets, micro plastics in cosmetics, toothpaste and now disposable masks from the Covid-19 pandemic. While plastics production is just over a century old, it was ramped up after World War II, transforming medicine, space travel, and safety gear (helmets, incubators, and clean drinking water equipment) (Parker, 2019). However, there has been an exponential increase in production from 2.3 million tonnes in 1950 to 448 million tonnes by 2015 (Parker, 2050). This skyrocketing production and poor waste management have led to between 4 million and 12 million metric tonnes of plastic entering the ocean each year; visually, this is enough to cover every foot of coastline on the planet, projected to triple its coastline cover over the next 20 years (Ocean Society, 2022). Now plastics are found in undesirable places,such as in drinking water, rivers, oceans, soil, inside aquatic animals and impacting our environment, human health, and the health of other living organisms.

As such, plastic has become one of the most pressing environmental threats of our time. The rapidly increasing production rates of plastic only exacerbate the harmful effects and yet never-ending presence. Plastic is not only a climate change accelerator throughout its life cycle, but it is also responsible for nearly CAD 8 billion in lost economic opportunity for Canada in 2019 (Wuennenberg and Tan, 2019). Years of research and remediation efforts have highlighted the intersectional multidisciplinary complex nature of the plastic problem that requires solutions beyond

[1] McMaster University.
[2] Åbo Akademi University.
[3] Thammasat University.
* Corresponding author: velmaigrover@yahoo.com

waste capture. Emissions attributed to the life cycle of plastic are partly related to the upstream emissions from extraction and refinement of oil and gas derivatives feedstock as well as manufacturing processes of resin and additive plastic products; also partly resulting from downstream emissions from end-of-life management practices, be it incineration, landfill, or faulty recycling and afterlife breakdown events (Hamilton et al., 2019). The low economic cost of fossil-based feedstock, its familiar functional capabilities,and its availability are foundational in its ubiquitous utilization in the production of over 380 million tons annually. If continued with no change, the production of plastic is projected to triple by 2050 (Ellen Macarthur Foundation, 2016), meaning over 56 gigatons of annual global carbon emissions (Hamilton et al., 2019).

The numbers tell a dire story of our relationship with plastics, e.g., every hour Americans throw away 2.5 million plastic bottles, totaling enough per year to circumnavigate the earth four times (MassDep, 2022). These plastics end up in undesirable places such as in drinking water sources, oceans, inside aquatic animals, impacting our environment, human health and the health of the environment. Although a lot of attention has been focused on the marine environment (the sink for plastic waste), the focus is now also shifting to terrestrial environments and major pathways of plastics or microplastics in rivers or freshwater systems to oceans. This book is an attempt to reveal many of these issues in a more scientific way, discuss the impact of plastics on human health and the health of the environment, including aquatic animals, and finally look at some global policy responses and technical solutions. It also highlights a research agenda for future work on this important topic.

Road map

The book is divided into three sections: a general introduction/science section, case studies, and policies followed by a conclusion. This chapter is followed by an introductory chapter to plastics, "Driver, Trends and Fate of Plastics and Microplastics Occurrence in the Environment", in which the author Emenda Sembiring discusses the origin and types of plastics followed by some models of transport and fate of the microplastics that help in understanding the sources and distribution of plastics. This information helps the stakeholders/decision makers to develop solutions/policies to deal with plastic waste and microplastic problems in the environment.

Most of the plastic-related research focuses on biodiversity and ecosystem degradation and the impact of both plastic production and plastic waste, while the contribution of plastic to the cumulative climate-relevant trace gases is often unrecognized. The problem of conventional plastic feed stocks is the core issue of the linear economy and its associated climate change stressors. The third chapter in the book, "The Correlation between Plastic and Climate Change" by Negin Ficzkowski and Gail Krantzberg explores the life cycle impacts of plastics on cumulative greenhouse gas emissions. It then investigates the cyclical nature of plastic pollution and extreme weather events. The chapter also illustrates, in brief, the impacts of plastic pollutants on the largest carbon sinks on earth (oceans) and investigates the adverse effects on global food supplies. Finally, the authors discuss landfill gas emissions and other land-related complications as they relate to plastic.

Waste recycling, reuse, and other similar practices might slow down the amount of waste produced, but waste is still being produced; the focus is now shifting toward a circular economy where the idea is to eliminate waste. Instead of cradle-to-grave, the new designs should focus on cradle-to-cradle. The next chapter in the book, "Plastics and Circular Econom" focuses on circular economy principles and their application to the plastics sector. Authors, Greg Zilberbrant and Eric Kassee, discuss how valued products and services need to be redesigned to eliminate plastic pollution and waste. The authors also look at the problem of plastics that are destined for disposal or already in the natural environment and the need to recover it followed by a section on how policies must be established to incentivize actions that eliminate or recovers plastic waste and clearly disincentivizes the opposite.

Just like anywhere else, plastic pollution is a big problem across the Laurentian Great Lakes. Plastic waste that ends up in these lakes originates primarily as litter on land and streams, rivers that provide inputs to the lakes, including Hamilton Harbour. Steven Watts, Christine Bowen,and Chris McLaughlin in "Seeing Is Believing: Educational and Volunteer Programming to Address Plastic Pollution in Hamilton Harbour" discuss the importance of understanding 'plastic litter budgets' for the success of future management strategies to reduce plastic litter in surface water bodies and avoid the negative consequences of the waste on aquatic environments. The authors discuss how poor public understanding of the relationship between the personal usage of plastics and the growing problem of plastic waste reaching surface waters presents a significant challenge to reducing the plastic problem. However, using Hamilton Harbour as a case study, the authors also show that a variety of community engagement activities and programs can improve public understanding and consequently create opportunities to advance behavioral and policy changes. As discussed by authors in the chapter, Hamilton Harbour has improved in terms of access to the waterfront, better water quality, etc., resulting from efforts to engage the broader community with programs and activities that provide teaching opportunities and tangible experiences for the public. Although it is easier to work on issues that are readily seen and sometimes readily sniffed,such as algae blooms and sewage bypasses,issues that are not as tangible, such as the extent and impacts of microplastics, require more focused efforts to educate and engage community members. The authors describe how the Bay Area Restoration Council is working toward providing and translating scientific research on microplastics and improving understanding of microplastics, a problem that will require multiple avenues to eradicate.

An estimated 10,000 metric tonnes of plastic waste flows into the Great Lakes annually, which is primarily due to ineffective waste diversion and recovery. A suite of coordinated actions is required to address the systemic causes of plastic pollution and its impacts on our critical freshwater ecosystems. This chapter, "The Great Lakes Plastic Cleanup: An Effective Approach to Addressing Plastic Pollution in the Great Lakes", introduces the Great Lakes Plastic Cleanup, an initiative of Pollution Probe and the Council of the Great Lakes Region with support from a network of collaborators aimed at addressing plastic pollution through the use of innovative plastic capture technologies to collect plastic before it enters the lakes and cleans up that which has already found its way there. The Great Lakes Plastic Cleanup uniquely combines a focus on removing plastic from the environment, community outreach and engagement, and the collection of locally-relevant data that can further inform regulatory and policy decisions. Framed as a case study, the authors (Melissa De Young, Mark Fisher, and Christopher Hilkene) present initial data and program strategies in the context of exploring how to effectively address plastic pollution and identifying areas where further research and action are needed.

Plastics are extensively used by the aquaculture sector from packaging to the manufacturing of various equipment. A considerable portion of equipment and other materials that are used in aquaculture culminate in marine and other aquaculture environments as waste. A variety of aquatic organisms, across different trophic levels, can absorb or consume microplastics (MP) readily. The next chapter "Microplastics Sources, Contamination and Impacts on Aquaculture Organisms" summarizes the current status of research that has been carried out on the detection of MPs in aquaculture organisms; MP pollution in mussels, MP sources in an aquaculture environment, mechanisms of interactions with organisms, contamination and effects on biota. The authors considered how various aquaculture organisms, such as mussels, sourced from farmed, wild environments and markets from around the world were reported to be polluted with MPs. As explained by authors, H.U.E. Imasha and S. Babel, the annual dietary intake of MPs through European shellfish can be high as 11,000 MPs per year. These findings show that the consumption of seafood can be a direct way of human exposure to MPs. Exposure of biota to MPs results in adverse physical damages and cellular and molecular level toxic effects, sometimes even leading to death. MPs' toxicity experiments on humans are still at an infancy level, making it hard to accurately determine the adverse effects on human health. This chapter argues for the necessity of more research regarding the toxicological effects of MP pollution on human health, which is still at nascent levels.

Yubraj Dahal and Sandhya Babel in "Occurrence and Source of Microplastics Contamination in Drinking Water and Subsequent Performance of Water Treatment Plants"present an overview of the occurrence and source of MPs in drinking water (tap water and bottled water). The abundance of the small-sized MPs has been observed to be considerably higher in bottled and treated water compared to larger plastic fractions. The authors also discuss the efficacy of drinking water treatment plants in removing the MPs.

The next section of the book focuses on what is being done in terms of governance and policy aspects to control/regulate plastic waste. The first chapter of this section is a case study of Canada, "Regulating the Export of Plastic Waste in Canada", by Ryder LeBlanc and Gail Krantzberg. Despite Canada's municipal recycling programs and the best intentions of its environmentally conscious citizens, plastic waste continues to be a serious issue both at home and internationally. News coverage in recent years has revealed how much waste has been shipped abroad, especially to Southeast Asia, where it continues to cause widespread pollution and threaten human health. In January 2021, Canada agreed to amendments to the Basel Convention, which made it nearly impossible to ship plastic waste overseas. The United States (US), though a signatory to this convention, has not ratified it and the plastic waste trade between Canada and the US lacks the regulations required to prevent the US from acting as a loophole for Canadian exporters. Therefore, a policy gap exists in this area. Based on current political trends in the US, it is unlikely that they will ratify the Basel Convention or support any legally binding agreements with Canada, so it is recommended that Canada revoke its agreement with the U.S. to ensure it can meet its international and domestic commitments on plastic waste.

The advent of plastic addressed many immediate needs for human society, incrementally creating a culture wherein eliminating it from the market is difficult to envision. Most of the efforts to deal with the adverse effects are focused on replacing its feedstock rather than rethinking the underlying design. As discussed by NeginFiczkowski and Gail Krantzberg in, "Nature Based Solutions to Address the Plastics Problem: Biomimicry", there has been a shift toward using renewable biomass sources, such as corn, cassava, sugar beet, sugar cane, or cellulose, that offer potentials for reducing the overall environmental impacts of the produced waste. The increasing appetite for solutions in this category has created a market that is nearly saturated with what is commonly known as 'bioplastics' which, contrary to popular belief, does not inherently address the emissions or pollution issues attributed to conventional fossil-based plastic. From a design standpoint, the pattern continues to follow the linear model of take-make-use-dispose, resulting in burden shifting throughout the life cycle of these products. In the case of the plastic crisis, any effective solution should simultaneously consider the environmental and climate costs resulting from systemic complexities and failures of the plastic industry over the past century. This chapter investigates the opportunities this shift provides for Canada, the existing barriers, and the changes required within the Canadian landscape to allow for such innovative ventures to become mainstream.

In line with the international scenario, microplastic research in India has also gained momentum during the last few years, as evidenced by the increasing number of publications. In this context, in the next chapter "Microplastic Research in India: Current Status and Future Perspective" authors Arunbabu V. and E.V. Ramasamy review the scientific data related to microplastic research in India. This chapter summarizes the data on microplastics from various environmental matrices, such as marine, freshwater, and terrestrial. Microplastics in biota and food items destined for human consumption were also given due importance. The contaminants commonly adsorbed on the surface of microplastics and the additives used in plastics are discussed with respect to their impact on living organisms. The authors have examined the existing policies and regulations in India and have proposed some solutions to control microplastics. The chapter also identifies the major challenges facing microplastic research in India and identifies the priority areas for future research.

Just like anywhere else in the world, daily living in South Korea also involves the use of a lot of plastic products, such as straws, cups and drinking bottles, and meal boxes and parcels wrapped in plastics ordered on line. The increasing use of plastic (leading to more waste) and the ban on exporting this waste to China and other Southeastern countries have led to rising trash mountains in the countryside. This has also led the legal enforcement in the country to become more stringent. The resulting increase in plastic waste in the environment compelled the Korean government to commit to cutting plastic waste by 2030. Youngjin Choi in, "Forging Plastic Governance: Addressing Acute Plastic Pollution in Korea", examines plastic use and suggests solutions on how to reduce plastic waste, such as the use of biodegradable plastics, while demonstrating environmental politics and forging multilayered plastic governance on plastic use and recycling in South Korea.

Since the plastics problem (especially in oceans) is global, solutions will require a multifaceted approach with cooperation at all levels, including the State-State level. International law provides one such framework which can enable state-state cooperation on issues of international concern, in this case, both the law of the sea and the law of international watercourses. Mostly 10 river systems are blamed for transporting plastics into oceans and eight of these are in Asia; each of which flows partially through China's territory. Four of them are trans boundary and therefore fall under international law. In the next chapter "The Role of International Water Law in the Management of Marine Plastic Pollution: The Case of China and its Trans boundary Rivers", David J. Devlaeminck explores the role of international law in the governance of marine plastics from rivers with a focus on China and its neighbors, highlighting pathways for China, which sent shock waves through the plastics industry with a ban on plastic waste imports in 2018 to play a leading role in combatting marine plastic in regards to its transboundary rivers.

There are various policy tools that can be used to combat plastic pollution. Nanuli Silagazde and Savitri Jetoo explore one such governance tool, referendum, in "The Plastic Vote: Referendum as a Governance Tool to Combat Plastic Pollution". Direct democracy is a vital element of governance in the United States that has been applied to resolve a wide range of policies. In recent decades over a hundred environmental issues have been put on popular vote addressing issues,such as energy, forests, natural resources and water. Despite the widespread use of referendums in this domain, there have been only two popular votes on plastic—in the state of California in 2016 and the city of Seattle in 2009. Both referendums were sponsored by the industry to overturn the previously imposed ban on single-use plastics with two contrasting outcomes. This chapter investigates why these referendums occurred in the first place, the narrative used in the campaigns and other factors contributing to their adoption or rejection by the public. Most importantly, it examines whether referendums can be viewed as an effective tool of governance in combatting the problem of plastic pollution. All these aspects gain additional weight amid COVID-19 since hard-won plastic bag bans have been suspended in various communities and the plastics industry is seizing the moment and lobbying for further changes in the legislation.

Considering the current global plastic waste situation, there are two urgent issues to be addressed: one is to reduce the volume of uncontrolled or mismanaged waste streams going into the water bodies (including oceans) and the other is to increase the level of recycling (UNEP 2021). There is also a need for a legally binding global treaty. In the final chapter of the book, "Plastic Pollution Treaty: Way forward", Neha Jununkar looks at some of the existing treaties, such as the Basel Treaty and its amendment, and discusses the most recent treaty to deal with plastic waste, "End Pollution Treaty", its draft form and the way forward to a better future.

Conclusion

As can be seen from the discussion above, plastic waste is a complex problem and needs to be addressed more holistically. This includes looking at better designs, more sustainable feedstock,and partnerships between various stakeholders, including the production industry, packaging industry, users, academicians/researchers, policymakers, and waste management industry. The idea is to have improved plastic design and products to facilitate reuse, repair, and recycling as suggested in the "European Strategy for Plastics in Circular Economy".

References

MassDep. 2022. News. Plastics. Presented by the Commonwealth of Massachusetts Department of Environmental Protection. Accessed online on Aug 16, 2022 at 8:30 am at (https://thegreenteam.org/recycling-facts/).

Ocean Society. 2022. The Ocean Plastic Pollution Problem. The Ocean Society. Accessed online on Aug 16, 2022, at 10:00 am at (https://www.oceanicsociety.org/resources/7-ways-to-reduce-ocean-plastic-pollution-today/).

Parker, L. 2019. The World's plastic pollution crisis explained. National Geographic June 7, 2019. Accessed online Aug 16, 2022 at 9:57 am at(https://www.nationalgeographic.com/environment/article/plastic-pollution).

Chapter 2

Driver, Trends and Fate of Plastics and Micro Plastics Occurrence in the Environment

Emenda Sembiring

BACKGROUND

Plastics and microplastics (MPs) have become one of the world's significant environmental issues in this decade; this condition occurs because of the massive usage of plastic-based products. Plastics are not only widely distributed but are also persistent in the environment. Plastics and MP sources are mostly diverted from the land and are transported into the ocean through the freshwater system. A comprehensive understanding of the source, fate, transport of plastics, and MPs is needed. Models of transport and fate of the MPs can be complementary information to understand the source and distribution of this polymer and help the stakeholders to overcome the plastic waste and MPs problem in the environment.

Before the definition of modern plastic, the term 'plastics' referred to anything that could be easily molded; the name is derived from the Latin *plasticus* and the ancient Greek *plastikos*. Recently, plastic materials have frequently been described by their physical qualities, owing to their lightweight, persistence, and durability as well as their capacity to efficiently insulate heat and electricity (Thompson et al., 2009).

Plastics are polymer compounds that are made from naphtha. Naphtha is a product of distilled crude oil to form a simpler molecule structure, like ethylene and propylene. Then the molecules are chemically bound to form a polymer, such as polyethylene and polypropylene, which is called synthetic resin. Before it can be used to form plastic products, it needs to be melted and added to some additives to

Air and Waste Management Research Group, Faculty of Civil and Environmental Engineering, Institute Technology Bandung, Jalan Ganesha 10, Bandung, Indonesia.
Email: emenda@itb.ac.id

form a pellet. From this point onward the pellet can be called plastic. The plasticscan then be shipped to manufacture to form a plastic product.

Plastics are defined chemically as synthetic organic polymers which are generated from the polymerization of monomers taken from crude oil or natural gas. Celluloid is often regarded as the first man-made plastic substance, even though it is composed entirely of nitrocellulose. Currently, the type of plastics includes bio-based plastics. Bio-based polymers differ from synthetic organic polymers only in their isotopic made-up of oil-based polymers. Even if this bio-based polymer is biodegradable, it should still be classified as plastic.

Plastics are engineered to be persistent and are designed as durable materials so that they will degrade very slowly up to a thousand years. If plastic consumption increases overtime, eventually there will be a need for more space to dispose of plastic waste. Burning plastic is not recommended as it will emit toxic fumes. Despite all benefits of using plastic materials, the manufacturing of plastics often creates large quantities of chemical pollutants as plastics production consumes 4% of crude oil, and it still consumes nonrenewable resources.

Plastic mass manufacture began in the 1950s, and it has risen 200-fold since then. Now, the yearly worldwide plastic output is projected to be 300 million tons (Napper et al., 2015), while total global production has reached 8.3 billion metric tons (Geyer et al., 2017). Over half of all plastic is used for single-use disposable products, mostly packaging (Xanthos and Walker, 2017). The plastics industry uses resin identification codes (RICs) to define and classify polymers; this categorization enables more efficient plastic recycling. This method was pioneered by the Society of the Plastics Industry (SPI) (The Plastics Industry Trade Association 2015). Figure 1 depicts the many types of plastics classified by their code and polymer composition.

1 PET	2 HDPE	3 PVC	4 LDPE	5 PP	6 PS	7 Others
Poly-ethylene Tetrap-thalate	High-density poly-ethylene	Poly-vinyl Chloride	Low-density poly-ethylene	Poly-prophylene	Poly-styrene	Other
Water bottles, jars, and caps	Shampoo bottles andgroceries plastics	Cleaning product and sheeting	Bread, bag, and plasticfilms	Yogurt cups, hangers, and straws	Take away packaging and electronic packaging	Baby bottles, Nylon, and CDs

Fig. 1. Plastic Category.

The types of plastics commonly used for daily activities and industry are generally described as follows.

1. PET is a polymer composed of linear polyesters (PES). Polyester is a very versatile plastic polymer with the same ester bond as vinyl chloride (–COO). These bonds account for a very little portion of the molecule's chemical structure. PET was patented in 1941, even though polyester resins were first synthesized in 1883. PET is the fundamental constituent of synthetic polyester fibers; therefore, it is widely used in clothes, ropes, and carpet fibers. PET is used in almost all mineral water bottles.
2. Polyethylene (PE) is the simplest organic polymer with a simple linear polymer. PE was initially marketed in 1933 and has been extensively used ever since owing to its cheap cost, superior electrical insulating characteristics, high chemical resistance, simplicity of manufacturing, high durability, flexibility, and transparency in thin film forms. Additionally, high-density polyethylene (HDPE) is extensively utilized in a variety of applications, including insulating electrical equipment and bottles. Meanwhile, low-density polyethylene (LDPE) bags are extensively utilized.
3. Polyvinyl chloride (PVC) is a vinyl chloride polymer that is synthesized through a copper catalyst reaction with oxygen and hydrogen chloride. The method of vinyl chloride polymerization was developed in 1872. PVC is a colorless, rigid substance that is brittle and unstable. As a result, additional chemicals are required to convert them into useable plastic materials. These polymers are then utilized in a variety of everyday and industrial applications, including pipes.
4. Polypropylene (PP) is formed when propylene is polymerized; this method was discovered in 1954 but commercial usage began in 1957. PP is typically a durable material. Additionally, this substance has the lowest density of any plastic. Plasticized polypropylene is extensively used in packaging, textile fibers (textiles, ropes, and carpets), and films.
5. Since the early 1930s, polystyrene (PS) has been marketed. This is because this type of plastic is inexpensive to manufacture, is transparent, has excellent electrical insulating qualities, and is stiff and simple to shape. These aromatic polymers are available in solid or foamed forms and are utilized in a broad range of applications, including packaging, food containers, and construction materials.
6. Other kinds of plastic (O), this category includes plastics produced from a variety of sources that are not classified as other types of plastic throughout the codification process. Examples of plastics classified as "O" include polyamide (PA) and polycarbonate (PC) as well as other kinds of plastic that have seen increased usage in recent years, such as PUR or polyurethane.

Driver and Trends

There are several factors that drive the increase in plastic usage. The drivers can be divided into two types, namely intrinsic property, and external influences. The intrinsic property relates to the beneficial property of plastics, such as versatility, lighter than other substitutes, durability, ability to tailor to meet specific technical needs, resistance to some chemicals, especially water, and to some extent good safety and hygiene property for food package and has excellent thermal insulation. It can be predicted that the intrinsic property will only explain the gradual pattern of consumption, not the dramatic increase in consumption. Currently, most countries consume plastics 20 times higher than in the past 50 years ago (Waste Watch, 2006). The presumption is that external influences, such as economic growth, lifestyle, urbanization,and technological process contribute more to the pattern of the dramatic increase.

Economic growth is the major factor to boost plastics usage. By using the GDP as an indicator of economic growth, it is found that the higher the GDP per capita, the more plastic consumption per capita (Fig. 2) (Ritchie, 2018). The other concern related to economic growth is the rise of the middle-class groups in the Asia Pacific region. The young middle-class population will be the driving force in Asian countries. The arrival of a billion more middle-class consumers will speed up the shift of East Asian economies from focusing on export-led development to more consumption-driven growth (McKinsey Report, 2007). The rise of middle class in China and India is most likely to enjoy the same lifestyle as in developed countries (McKinsey Report, 2007). This consumption-hungry group has the financial ability and desire to spend their growing income on luxurious products, not only to survive with enough food but also on high-quality food, a comfortable living environment, and personal beauty. This middle-class group asked for a more sophisticated lifestyle and quality of life (QoL).Unfortunately, the increase in QoL usually relates to more consumption and a change in lifestyle to be more practical. For example, nowadays in most Asian countries, more urban societies prefer to shop at the supermarket where most of the food is wrapped in plastics, instead of going to a traditional market which hardly uses plastic materials to wrap food. It is convenient, but it contributes to more plastics package used in the environment.

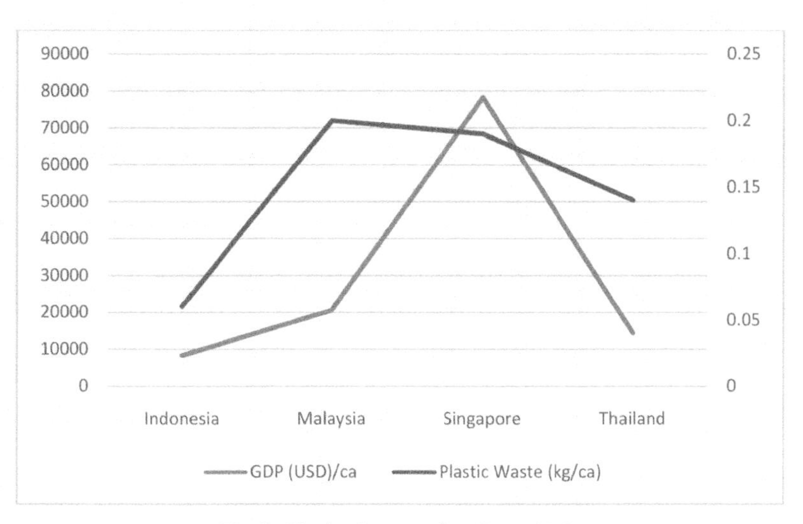

Fig. 2. Plastics Consumption Versus GDP.

The improvement in financial ability and the availability of a wider range of products led to a change in the consumption pattern. In other words, people's lifestyle has changed and the quality of life (QoL) has improved. China, for example, follows this pattern. In 1978, China was dominated by the agricultural sector and since 1842, it shifted to be a more industrial country (Guan and Hubacek, 2004). This period showed only a slight change in people's demands and lifestyles. On the contrary, since 1978, because of the policy and infrastructure support, economic development has increased dramatically. The production of a wider range of products and the high purchasing power of consumers led to dramatic changes in lifestyle in China. Fig. 3 (Zhixin et al., 2002) shows that increase in purchasing power will increase the expenditure on food and clothes. Again,more consumption of food and clothes will contribute toward more packaging waste, especially if it is followed by changes in lifestyle.

Most Asian countries have transformed their economy from depending on the agriculture sector to the industrial sector. Shifting structure affects the occupational of people. More rural peasants come to the urban area seeking jobs. It is expected that more rapid urbanization will occur in the next 25 years in most Asian countries. More people will live in urban cities, which have their lifestyles and characteristics. As a consequence, the urban lifestyle consumes more food and clothes than the rural lifestyle (Fig. 4) (Zhixin et al., 2002). According to Huang and Bouis (2001), urbanization has significant influences on food demand. Changes in marketing systems and occupational changes are closely linked with increasing GNP per capita and also may influence the demand for food. Similarly, Korea demonstrated that structural changes in food demand because of the changing income has been a significant factor for the rapid changes in dietary patterns in East Asia over the past three decades (Huang and Boui, 2001).

Based on the three above-mentioned paragraphs, the drivers of both intrinsic factors and external properties change the consumption pattern. People consume more, especially food in this case. The relation of change in food consumption habit and plastics waste relate to the current trend in how manufacturers, producers, and retailers use wrapping materials. The current trend of technology in food processing leads to more plastics that are used as packaging materials. More processed food, ready-made food, and fast food are readily available in the market. Because of the availability of these convenient foods and changes in the food consumption habit,

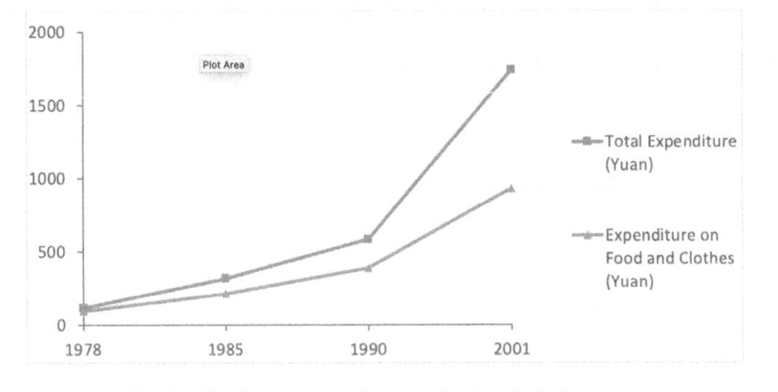

Fig. 3. The Rural expenditure on food and cloth pattern.

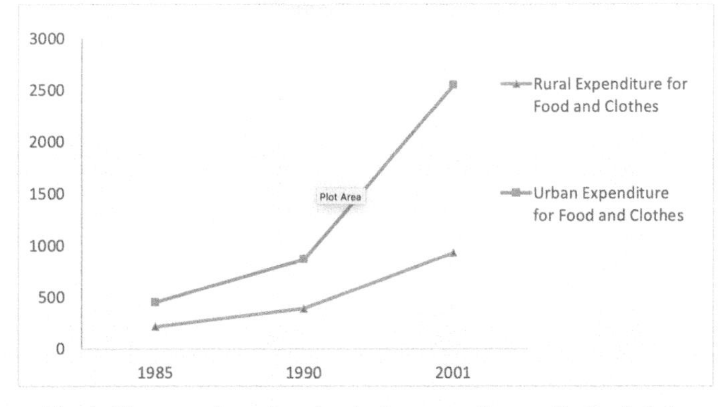

Fig. 4. The comparison of rural and urban expenditure on food and cloth.

more and more packaging waste will be thrown away in the environment. A similar case occurred during the Covid-19 pandemic that led to changes in consumer behavior toward the online market. More buying from the online marketplace means more packaging waste will be generated.

In addition, the improvement of injection molding and extrusion machines and high-speed food processing machinery increases the possibility of various foods being processed and wrapped with packaging material, especially plastics. For example, an increase in the technology and characteristics of plastic film that film can balance the respiration of oxygen (O_2) and carbon dioxide (CO_2) so that it will increase the food's life span. The food wrapper, especially for vegetables and fruits, can exchange gases better and in turn, it will enhance their life span. If this technology is applied, the food life span can be increased. However, as a consequence, more plastic waste will introduce into the environment.

Global Plastic Debris and MP Models in The Ocean Studies

Plastic and MPs have become one of the world's biggest environmental issues in this decade. This condition occurs because of the massive usage of plastic-based products being widely distributed, and this polymer is persistent in the environment. In 2015, a global model shows countries that diverted plastic waste into the ocean. Countries such as China, Indonesia,and the Philippines were identified as three major countries responsible for releasing plastics wastes into the ocean (Jambeck et al., 2015). Most of the studies show that the source of plastic debris and MPs were mainly coming from land. A current study on global MPs' mass balance shows hundred to thousands of plastic debris in the ocean (Lebreton et al., 2019). It takes time before the plastics emission from the land bases can accumulate in the ocean. Most of the global plastics debris model input was based on population density, level of urbanization and industrialization, and source of plastics/MPs (GESAMP, 2015; Jambeck, 2015). The current model has been refined to include hydrological information, such as topography information, rainfall, and catchment area (Lebreton, 2016). The river located in Asia accounted for 67% of the global total of plastic debris in the ocean.

Table 1. Global Plastics Waste Debris.

Study	Generation of Plastics Debris	Remarks
Jambeck et al., 2015	4.8–12.7 million MT/year	Population density and economic status are the main input of the model. Population size and quality of solid waste management determine the country's contribution to plastic and marine debris.
Lebreton et al., 2016	1.15–2.41 million MT/year from the river to the ocean.	Runoff plays a significant role in plastic waste emissions into the ocean. Significant sources are wastewater and tributaries, while the significant sink is weirs.
Lebreton et al., 2019	There is 65% of global plastics production. Clarifies the global prediction gap; hundreds to thousands of MT plastics have been accumulated in the ocean.	Mitigation: (1) reduces emissions and (2) removes the accumulation in the environment.

A typical model based on Lebreton et al. (2019) not only shows the source of waste based on population density but also includes the mismanaged waste from solid waste facilities (PPC). The model improvement now also includes the hydrological data.

Plastics in the ocean are present not only in the form of debris but also in the form of MPs. MPs are generally defined as synthetic polymers less than 5 mm (Klein et al., 2015; Thompson et al., 2009). These pollutants can be distributed by various routes of transport, including river flow, wastewater discharge, wind, and runoff (Gall and Thompson, 2015). In addition, MPs can also come from various sources but in general, the main source of these MPs is human activities in terrestrial and aquatic environments. The terrestrial environment is the main source, based on the percentage of MPs from this source, which is around 80%, while marine activities contribute only about 20% (Andrady, 2011). Based on the process of formation of MPs, it can be divided into two types. The first is primary MPs, the plastics that are already in micro-size and usually found in cleaning and cosmetic products (Cole et al., 2011b). The second source is secondary MP, and this type is formed from bigger plastics in the environment and fragmented into micro-sized plastic (Browne et al., 2011).

This pollutant is widely distributed, and MPs have been found almost in every freshwater system in the world (Eerkes-Medrano et al., 2015) and even in Antarctica (Isobe et al., 2017; Lusher et al., 2015) and spreads from the surface water to the sediment (van Cauwenberghe et al., 2013). The very wide distribution pattern is strongly influenced by the flow and circulation of water, yet the characteristics of MPs, such as density, shape, and size, also affect the distribution of MPs (Ng and Obbard, 2006). A recent study shows that the freshwater system in Southeast Asia is also polluted by MPs. A review of marine MPs' abundance in the ASEAN ocean was conducted and showed that almost all countries in this region are polluted by this polymer (Curren et al., 2021). MPs are also found in the river in ASEAN region; for example, in the Chao Phraya River (Thailand) (Ta and Babel, 2020) and Citarum

River (Sembiring et al., 2020), a recent study found that this contaminant polluted one of the biggest reservoirs in Indonesia (Ramadan and Sembiring, 2020).

MP is found in all areas of the freshwater system based on this condition, and there is a need to systematically understand the fate of transportation of MPs in the freshwater system. The current study of MPs' abundance and characteristics can become the baseline for the development of MP models.

MPs' Conceptual Model for the Fate and Transport in Freshwater System

The abundance of MPs in freshwater systems originates from various types of sources, such as effluent from waste water treatment plants, direct disposal of plastic into water bodies, storm water runoff, plastic leakage from solid waste management, and atmospheric deposition (Fig. 1) (Lu et al., 2021). When primary MPs enter the water bodies,density, polymer type (Browne et al., 2011), and shape (He et al., 2019; Khatmullina and Isachenko, 2017) impact the dispersion of MPs. Secondary MPs will have different dispersion mechanisms due to the occurrence of these MPs which originate from the fragmentation of bigger plastic. The fragmentation process of MP in the environment is affected by the physical abrasion photo degradation and biodegradation by weakening the surface of MPs (Song et al., 2017; Zhu et al., 2020). Transport and fate processes in the freshwater system have been affected by many processes; for example, in the water, advection, dispersion, and diffusion will make the MPs' distribution varied (Daily and Hoffman, 2020; Kooi et al., 2017).

Based on the source, MPs can be divided into primary and secondary sources. Primary sources are micro-sized plastic polymers which are found in cleaning and cosmetic products, such as scrubbers as well as pellets that are produced for use as raw materials for plastic production (Cole et al., 2011a; Fendall and Sewell, 2009). The secondary source of MPs is in the form of fibers or fragments resulting from the fragmentation of larger plastics (Browne et al., 2011). MPs from these secondary sources are often associated with areas of high population density (Ballent et al., 2013; Desforges et al., 2014). In addition, according to (Hidalgo-Ruz et al., 2012) secondary sources are the largest source that causes an abundance of MPs in the marine environment (Fig. 5).

MP transports in lotic ecosystems, such as reservoirs and lakes,which have a different mechanism to the MP model in the lentic ecosystem, such as rivers. The mechanism in this environment tends to distribute slowly and the mechanism, such as sinking and wind-affected movements, tends to make this ecosystem different from lotic ecosystems, such as rivers. Furthermore, mixing processes in lakes are often complex to model because of vertical stratification and the lake geometry or bathymetry. Moreover, in rivers, the flow is substantially faster than in lakes and will contribute to the advection flow component value which is typically above the dispersion value (Ji, 2008).

Several modeling efforts are conducted in recent years, the model includes an emission-based model, a global model, a spatial-temporal model, and hydrodynamic models in lakes or reservoirs.

Fig. 5. Plastics and MPs' Transport and Fate Scheme.

Emission-based model

The emission model was conducted in several studies; for example, the study in the Netherlands (van Wezel et al., 2016) estimated the MPs' emission from the WWTP effluent. Emissions were calculated based on the known usage of MPs in cosmetics and personal care items, cleaning agents, and paints and coatings. Data on product use, market penetration, and MP content in the product were gathered for each product category. During the wastewater treatment process, it was calculated that between 40 and 96% of the MPs would be retained by the WWTP. The predicted concentration of MPs in a WWTP effluent was calculated as the product of the MPs' concentration in a product, the daily usage of that product, the fraction of MPs removed during wastewater treatment and the market penetration of the products that is divided by the volume of wastewater produced by the model. When 'big and heavy' particles, i.e., particles with a relatively high density, large size, and large volume were supposed to be observed, the model best-matched data. The observed concentrations, on the other hand, comprise both primary and secondary plastics, but the model only included primary plastics which might result in a higher error to the claimed validation.

Moreover, the recent study about emissions conducted by Kawecki and Nowack (2020) utilized geographical information on land-use statistics, traffic and population densities, wastewater treatment facilities, and combined sewer overflows as proxies to regionalize macroplastic (MaP) and MP emissions for soil, freshwater, and air. The emission data for seven commonly used polymers (low-density polyethylene, high-density polyethylene (HDPE), polypropylene (PP), polystyrene, expanded polystyrene, polyvinyl chloride, and polyethylene-terephthalate) were used to create high-resolution maps of emissions for MPs and macroplastics.

Spatiotemporally explicit models in rivers

In recent years, the development of spatiotemporal MP models in the river took into consideration the river characteristics and particle process in the surface water.

The first model developed by Besseling et al. (2017) for Dommel River used the NanoDUFLOW hydrological model that comprises an advective method of transport, homo aggregation and heteroaggregation of particles, biofouling, sedimentation/resuspension, plastic degradation, and sedimentary burial. The model is not yet formally validated due to the limited monitoring data. This model is the continuation of a model by Quik et al. (2015) for nanomaterials in neutral waters.

The second study of spatiotemporal also conducted by Nizzetto et al. (2016) was entirely theoretical as no empirical data on emissions and concentrations of MPs were available. The model is based on the current multimedia hydrobiogeo chemical model, INCA contaminants with a transport sediment module, a rainfall-runoff module, and the option of adding inputs of direct effluents, such as the WWTPs. It is a lumped model because it assumes a uniform distribution of rainfall and temperature. Surface runoff and effluent inputs, as well as reentry into the system via resuspension, were all factored into the model. The amount of MP accessible for mobilization, the transport capacity of the overland flow for both MPs and sediments, and the detachment of plastics through splash erosion and flow erosion all had a role in whether particles were carried by surface runoff.

The latest MPs' model used estimation in tire wear particles and transportation to the Göta River, Sweden, uses hydrodynamic modeling (Bondelind et al., 2020). The goal of this study was to see how the size and density of tire wear particles in road run-off affected their fate in Sweden's Göta River. MIKE 3 FM software was used to create a model of the river, Sweden's largest river, going through Gothenburg (Sweden's second-largest city) and out to the sea. The following MP parameters were defined using literature data: storm water concentrations, prevalent particle sizes, the density of MP typically found in road run-off and settling velocities. The vertical water density gradient generated by salt water from the Kattegat strait has a significant impact on the river's mixing processes and MPs' concentrations. While most MP with a greater density and bigger size settle in the river, smaller MP with a density of less than 1.0 g/cm^3 do not and therefore reach the Kattegat strait and marine region.

Spatiotemporally explicit models in lakes

The recent study of MP modeling, the three-dimensional transport, and distribution of multiple MP polymer types, in Lake Erie was conducted by Daily and Hoffman (2020). The model simulated the motion of plastic pollution in Lake Erie affected by advection, density-driven sinking, and turbulent mixing using a Lagrangian transport model to explore the distribution of plastic in the water column and sediment. The model tracks particles that hit the bottom to represent deposition, and it includes nine polymer types that account for over 75% of predicted global plastic waste. To calibrate the model and derive estimates for the mass of plastic in the lake volume and the flux of plastic into the sediment, the model's spatial distributions are compared to surface samples. The mass estimate of 381 ± 204 metric tons is two orders of magnitude higher than earlier surface estimates, although it is still a small proportion of the annual inflow expected. The findings represent a step toward resolving the plastic mass balance in Lake Erie and gaining a better understanding of plastic transit into the sediment.

Plastic Leakage from Land to the Waterways

Plastics enter the marine environment through a number of routes, including river and atmospheric transmission, beach littering, and direct discharge at sea through aquaculture, shipping, and fishing (GESAMP Joint Group of Experts on the Scientific Aspects of Marine Environmental Protection, 2015). However, in comparison to marine-based sources, land-based sources are regarded to be the primary source of plastics in the seas (Jambeck et al., 2015). The origins of plastic waste in and around freshwater systems are directly connected to human activity since the amount of plastic in rivers is highly correlated with population density, urbanization, wastewater treatment, and waste management (Best, 2019). Additionally, landfills, the most prevalent method of disposing of solid waste, are the primary repository and disperser of plastic. MPs, including primary and secondary MPs, are also generated in landfills (He et al., 2019a). Numerous natural processes (e.g., rain and wind) contribute to the spread of landfilled MPs into surrounding ecosystems such as aquatic systems (Yadav et al., 2020).

Plastic waste enters river systems by natural transport mechanisms or direct dumping as a result of a lack of MSW or inappropriate community behavior. Once plastic enters the river, hydrological parameters like water level, flow velocity, and discharge impact plastic transport (van Emmerik et al., 2018). The fate of macroplastics that enter freshwater systems remains a major uncertainty in studies on river plastic transport. Generally, it is thought that all plastics found in rivers end up in the ocean. However, a substantial portion of plastic pollution (99%) is never discovered afloat in the seawater and is therefore deemed "lost" (Gigault et al., 2018; van Sebille et al., 2018). Recent research indicates that these plastics sunk under the water surface (Choy et al., 2019). Additionally, the polymer type, shape, and size of plastic are essential factors in the fate of plastics or MPs. It has been discovered that the characteristics of plastic have a significant effect on how plastics are distributed in the environment (Schwarz et al., 2019). Plastics' fates in freshwater systems are highly reliant on three processes: (1) transport, (2) accumulation, and (3) the degradation processes that all plastics develop in the environment. A secondary stream of micro- and nano-plastics is generated, while the macro-plastic degrades (Andrady, 2017).

According to Browne et al. (2011), plastic waste in the aquatic environment can be degraded by various processes, including biological, ultraviolet (UV), thermal, physical and thermo-oxidizing hydrolysis. For most types of LDPE and HDPE, polypropylene (PP) and nylon are degraded by a photo-oxidation process using UV-B light which is followed by thermal oxidation degradation. When plastic is exposed to UV-B light, chemical polymer bonds become unstable, resulting in structural damage. This degradation process might result in macromolecular particles

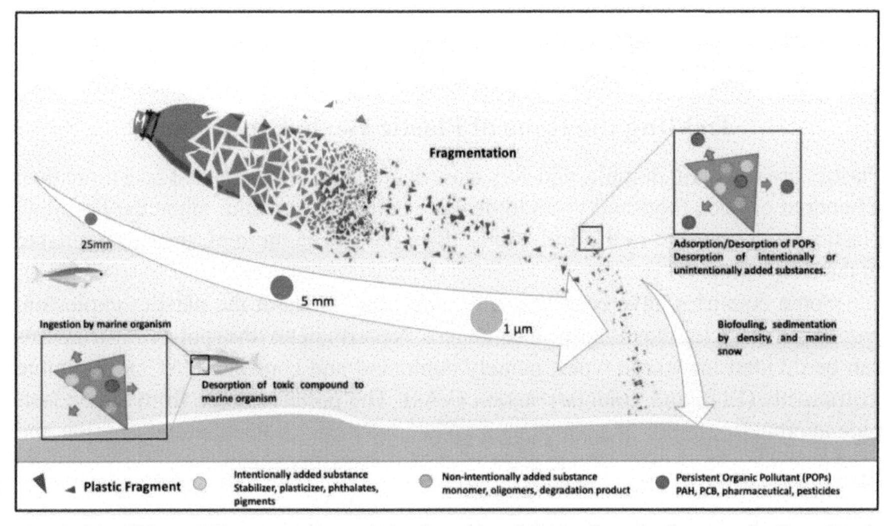

Fig. 6. The Effect of Fragmentation and the Sorption of Toxic Organic Compounds Onto Plastic Fragments on Marine Organisms.

degrading into micromolecules and even nanomolecules (Cole et al., 2011a; Galgani et al., 2013). The fragmentation process is shown in Fig. 6 (Masry et al., 2021).

Additionally, these fragments are often degraded biologically by the metabolism of particular bacteria and/or fungi which degrade the polymer structure via enzymes or secretory secretions. Moreover, oligomers and monomers will enter the cytoplasm where they will undergo mineralization, converting the plastic pieces into energy and structural components of the cell (carbon, nitrogen, phosphorus, sulfur, etc.) (Andrady, 2011). The process of plastic mineralization is predicted to take hundreds to thousands of years (Barnes et al., 2009). Pressure, friction, and stress contribute to the brittleness, fracture, or damage of MPs, such as fiber decay during the washing process, pressure forces on waves, and wind friction forces (Barnes et al., 2009).

Potential Solution to the Plastics Waste Problem

Based on the discussion above, it is clear how important waste management is, especially plastic waste management to prevent plastic waste leakage into the ocean (Lebreton et al., 2017; 2019; Best, 2019). Some plastics waste management practices include as following.

Plastic bag management

There are numerous examples of how plastic bag policy is applied in several countries with varying degrees of outputs/outcomes. The policy instruments range from the traditional approach [such as command and controls (CACs)], capital approaches[such as economic instruments (EIs)], or more related to ethics and behavior [such as voluntary action (VAs)]. The command-and-control instruments include product bans or fines on littering plastic bags, and the EIs include taxes and levies of products at producers, retailers, or consumers, whereas the voluntary actions include of promotion of alternative bags or following a certain code of conduct.

Tackling the Issue of Plastic Waste Explosion

Plastics are resistant, durable, and very slow degradable materials that take more than a hundred or even thousand years to degrade naturally. It means since the first time plastics were invented in the mid-1800s, there have been more plastics accumulated in the environment.

Some countries have been trying to tackle the issues of the plastics explosion. Numerous policies have been formulated. Nevertheless, the policy instruments can be divided into three types, namely command and controls (CACs), economic instruments (EIs), and voluntary action (VAs). The policies range from the end-of-pipe approach to more avoidance approaches,such as sustainable consumption and production.

Australia, for example,has implemented voluntary action since 1999. It has introduced the National Packaging Covenant that required the signatory parties and in this case between the government and the packaging supply chain to reduce the impact of consumer packaging including waste avoidance at the early stage of the product. Each party was required to submit the 3–5 years plan to meet the covenant obligation and contribute an annual fee that will use for funding the recycling project. Singapore also implemented voluntary action that asked packaging industries to sign the agreement on the promotion of the saving and recycling Act and 'The Singapore Packaging Agreement'. Both examples relate to EPRs as the producers are liable for the cost of managing their product at the end of their life cycle.

Most European countries applied the EIs, such as Austria, Denmark, Ireland, Netherlands, and the UK. In Austria, for example, producers, fillers, and importers are responsible for ensuring the collection of their used packaging or they can transfer the responsibility to a third party. In Denmark, the business sectors are responsible to separate their waste, and the local authority that ensures the implementation of a recovery program. Ireland introduced a levy on all plastic bags, except for those used for fresh products. The UK focused more on the possibility of a recyclable product as it aimed at meeting the target of recovery and recycling by means of tax and levy on products. Similarly,the Netherlands also focused on measures to reduce the quantity of packaging, for example, by using as little material as possible or using more recyclable content and avoiding the creation of litter.

The EIs usually affects either producers/business entity or consumers. Taxes can be collected from producers (brand owners, fillers, packers, and importers) or consumers. Product banned is the most prominent example of the CACs. Bangladesh, China, India, and Korea banned plastic bags as well. India, for example, banned the manufacture of plastic bags thinner than 20 microns in Bombay, whereas Bangladesh banned all polyethylene bags in the capital, Dhaka. Mainland China banned lightweight plastic bags thinner than 25 microns.

Most of the EIs are backed up by the CACs as the EIs work effectively if the regulatory standard and enforcement capacity are available. For example, Denmark also introduced a tax on packaging material, and it is regulated on the Consolidated Act No. 726 of 7 October 1998 (as amended by Act No. 912 of 16 December 1998, Section 9 of Act No. 380 of 2 June 1999, Section 23 of Act No. 165 of 15 March 2000, Section 13 of Act No. 1029 of 22 November 2000 and Act No. 1292 of 20 December 2000). The act describes the standard and application of packaging tax.

The results of instruments vary from country to country based on the socio-cultural, political, and capacity of enforcement of the country. In Asia, a strong CACs instrument is still used as the main measure to deal with the used plastic material. However, the effectiveness of the CACs instrument depends on the availability and capacity of the enforcement institution. Most of the EU countries set the CACs instrument to enforce the application of the EIs. The voluntary action may only be successful where environmental awareness is high, and the economic driver has met its target.

Most of the instruments are government driven at both national and local levels. On the contrary, some voluntary actions may be initiated by business entities. However, some voluntary actions may also be driven by the CACs. For example, in Indonesia, the retail business read the trend and prepare for the application of the new Act relating to solid waste management. In Act No 18/2008 of Indonesia Regulation, the extended producer responsibility is introduced, even though the application and the extent to which it should be followed are still in question. Most of the articles are too general to be applied as the operational details have not been regulated at the scale of *Peraturan Pemerintah, PP* (government regulation), *KepMen* (Ministry decree), and *Perda* (municipality or LAs regulation). It was stated that the detailed regulation will be introduced in 2009.

In article 14, Act No 18/2008 of Indonesia Regulation, every producer should put a label or information that relates to the reduction and waste management of products on packaging or products. In article 15, the producers must manage the packaging that they produce, especially if the packaging is difficult to degrade naturally. Based on Act no 18/2008, the producers must take responsibility if their products become waste. Article 40 describes that if any producers mismanage their waste or if they are not following the norm, standard, procedure, or criteria, the sanction includes jail for up to 10 years or a fine of up to 5 Billion Rupiah. The launch of Act 18/2008 is a little bit shocking to producers and retailers. Besides, there are some barriers that the retailers find difficult to apply Act No 18/2008: (1) no details application and sanction to the omission of the act; (2) competition among retailers that most retailers focus on and prefer the consumers' convenience; (3) consumers' lifestyle that prefer plastic bag instead of bringing their bags; (4) the possibility of the traditional market is not included in the act (Adiwijaya M, 2008). Recently, the Government of Indonesia, under the Ministry of Environment and Forestry, regulated the road map of waste reduction from producers (KepmenLHK P.75/MENLHK/SETJEN/KUM.1/10/2019). It shows how regulation can lead to more solid waste management system development.

Causal Evidence: A Case of EU Countries and Reflection of Indonesian Case

In order to see the future packaging waste management, especially plastics, this paper discusses the causal evidence of an additional policy based on the facts of EUs countries. The reason to pick the EU case as an analogy is that the author has observed in Asian developing countries, especially in the urban areas, their society follows the same pattern of consumption, even though the level of consumption is different. The goal of the policy is to achieve an effective policy to protect the environment. Packaging waste is a growing waste stream, even though the packaging directive has been in place for a decade in the EU. It shows that packaging waste has increased despite the goal of waste prevention. In the EU, the amount of packaging waste increased by 7% over the period 1997–2001. According to McGlade 2004, this trend is continuing if the business as usual is applied. Why has the amount of waste increased despite the application of the packaging directive? The general answer is that packaging waste is closely related to production and consumption. An effective policy to prevent more environmental impact should be related to decoupling waste generation from growth. A more specific factor in the EU is that a large percentage of packaging waste relates to the consumption of food. Consumer needs more pre-prepared foods which require more packaging. This trend is also observed in Asian developing countries, such as China (see again Fig. 4). This increasing trend is cause for concern because the generation of waste always has environmental impacts and represents a loss to society in terms of energy and materials.

All of the studies indeed show that recycling creates less impact on the environment than its disposal. But the fact remains that the environmental impact of packaging does not occur only in the waste handling but also in the production, transportation, and use of packaging. Therefore, the achievement of the recycling and recovery target is good for the environment. But how good is good? It is important to achieve more efficient policies to reduce the impact on the environment, such as addressing the broader objective of waste prevention. The other idea is that from an economic perspective, the marginal economic cost of increasing recycling is generally higher than what has been recycled (McGlade, 2004).

Reflecting on the EU case, it can be concluded that most developing countries, especially Indonesia in this case should propose additional policies besides the 3Rs strategy (in progress). The frog leap policy solely focuses on the national 3Rs strategy to be more production and consumption approach is justified.

References

Adiwijaya, M., *Peran Pemerintah, IndustriRitel, dan Masyarakat dalam Membatasi Penggunaan Kantong Plastik Sebgai Salah satuupayapelestarianlingkungan.* http://74.125.155.132/search?q=cache:4omcjaN0y8oJ:fportfolio.petra.ac.id/user_files/04-013/ARTIKEL-MICHAEL.ADIWIJAYA-UK.PETRA. doc+pengaruh+uundang+undang+no+18,+2008+buat+produsen+dan+retailandcd=2andhl=jaandct=clnkandgl=jp.

Andrady, A.L. 2011. Microplastics in the marine environment. *Marine Pollution Bulletin* 62(8): 1596–1605.

Andrady, A.L. 2017. The plastic in microplastics: A review. *Marine Pollution Bulletin* 119(1): 12–22. https://doi.org/10.1016/J.MARPOLBUL.2017.01.082

Ballent, A., Pando, S., Purser, A., Juliano, M.F. and Thomsen, L. 2013. Modelled transport of benthic marine microplastic pollution in the Nazaré Canyon. *Biogeosciences* 10(12): 7957–7970. https://doi.org/10.5194/bg-10-7957-2013.

Barnes, D.K.A., Galgani, F., Thompson, R.C. and Barlaz, M. 2009. Accumulation and fragmentation of plastic debris in global environments. *Philosophical Transactions of the Royal Society B: Biological Sciences* 364(1526): 1985–1998. https://doi.org/10.1098/rstb.2008.0205.

Besseling, E., Quik, J.T.K. and Sunand, M. Koelmans, A.A. 2017. Fate of nano- and microplastic in freshwater systems: A modeling study. *Environmental Pollution* 220: 540–548. https://doi.org/10.1016/j.envpol.2016.10.001

Best, J. 2019. Anthropogenic stresses on the world's big rivers. In *Nature Geoscience* (12(1): 7–21). Nature Publishing Group. https://doi.org/10.1038/s41561-018-0262-x.

Bondelind, M., Sokolova, E., Nguyen, A., Karlsson, D., Karlsson, A., and Björklund, K. 2020. Hydrodynamic modelling of traffic-related microplastics discharged with stormwater into the Göta River in Sweden. *Environmental Science and Pollution Research* 27(19): 24218–24230. https://doi.org/10.1007/s11356-020-08637-z.

Browne, M.A., Crump, P., Niven, S.J., Teuten, E., Tonkin, A., Galloway, T. and Thompson, R. 2011. Accumulation of microplastic on shorelines woldwide: Sources and sinks. *Environmental Science and Technology* 45(21): 9175–9179. https://doi.org/10.1021/ES201811S.

China statistics press, 2009. Retrieved at http://www.stats.gov.cn/english/statistical data/yearly data/YB2002e/ml/indexE.htm.

Choy, C.A., Robison, B.H., Gagne, T.O., Erwin, B., Firl, E., Halden, R.U., Hamilton, J.A., Katija, K., Lisin, S.E., Rolsky, C. and S. Van Houtan, K. 2019. The vertical distribution and biological transport of marine microplastics across the epipelagic and mesopelagic water column. *Scientific Reports* 9(1). https://doi.org/10.1038/s41598-019-44117-2.

Citizen information. 2008. Plastic bag environmental levy in Ireland. http://www.citizensinformation.ie/categories/environment/waste-management-and-recycling/plastic_bag_environmental_levy.

Cole, M., Lindeque, P., Halsband, C. and Galloway, T.S. 2011a. Microplastics as contaminants in the marine environment: A review. In *Marine Pollution Bulletin*. https://doi.org/10.1016/j.marpolbul.2011.09.025.

Cole, M., Lindeque, P., Halsband, C. and Galloway, T.S. 2011b. Microplastics as contaminants in the marine environment: A review. *Marine Pollution Bulletin* 62(12): 2588–2597.

Curren, E., Kuwahara, V.S., Yoshida, T. and Leong, S.C.Y. 2021. Marine microplastics in the ASEAN region: A review of the current state of knowledge. In *Environmental Pollution* (Vol. 288). Elsevier Ltd. https://doi.org/10.1016/j.envpol.2021.117776.

Daily, J. and Hoffman, M.J. 2020. Modeling the three-dimensional transport and distribution of multiple microplastic polymer types in Lake Erie. *Marine Pollution Bulletin* 154. https://doi.org/10.1016/j.marpolbul.2020.111024.

Desforges, J.P.W., Galbraith, M., Dangerfield, N. and Ross, P.S. 2014. Widespread distribution of microplastics in subsurface seawater in the NE Pacific Ocean. *Marine Pollution Bulletin* 79(1–2): 94–99.

Economic Instrument. 2008. Plastic bag.http://www.economicinstruments.com/index.php/solid-waste/article/188-

Eerkes-Medrano, D., Thompson, R.C. and Aldridge, D.C. 2015. Microplastics in freshwater systems: A review of the emerging threats, identification of knowledge gaps and prioritisation of research needs. *Water Research* 75: 63–82. https://doi.org/10.1016/J.WATRES.2015.02.012.

EU, European Union, European Parliament and council directive 94/62/EC on Packaging and packaging waste.

Fendall, L.S. and Sewell, M.A. 2009. Contributing to marine pollution by washing your face: Microplastics in facial cleansers. *Marine Pollution Bulletin* 58(8): 1225–1228. https://doi.org/10.1016/J.MARPOLBUL.2009.04.025.

Galgani, F., Hanke, G., Werner, S. and de Vrees, L. 2013. Marine litter within the European Marine Strategy Framework Directive. *ICES Journal of Marine Science* 70(6): 1055–1064. https://doi.org/10.1093/ICESJMS/FST122.

Gall, S.C. and Thompson, R.C. 2015. The impact of debris on marine life. *Marine Pollution Bulletin* 92(1–2): 170–179. https://doi.org/10.1016/J.MARPOLBUL.2014.12.041.

GESAMP Joint Group of Experts on the Scientific Aspects of Marine Environmental Protection. 2015. Sources, fate and effects of microplastics in the marine environment: a global assessment. *Reports and Studies GESAMP*, *90*, 96. issn:

Geyer, R., Jambeck, J.R. and Law, K.L. 2017. *Production, Use, and Fate of all Plastics Ever Made.* https://www.science.org.

Gigault, J., Halle, A. ter, Baudrimont, M., Pascal, P.Y., Gauffre, F., Phi, T.L., elHadri, H., Grassl, B. and Reynaud, S. 2018. Current opinion: What is a nanoplastic? *Environmental Pollution* 235: 1030–1034. https://doi.org/10.1016/j.envpol.2018.01.024.

Guan, D. and Hubacek, K. 2004. Lifestyle changes and its influence on energy and water consumption in China, University of Leeds. Retrieved at http://homepages.see.leeds.ac.uk/~leckh/leeds04/6.5final-gdb-march%20conference.pdf

He, P., Chen, L., Shao, L., Zhang, H. and Lü, F. 2019a. Municipal solid waste (MSW) landfill: A source of microplastics? -Evidence of microplastics in landfill leachate. *Water Research* 159: 38–45. https://doi.org/10.1016/J.WATRES.2019.04.060.

He, P., Chen, L., Shao, L., Zhang, H. and Lü, F. 2019b. Municipal solid waste (MSW) landfill: A source of microplastics? -Evidence of microplastics in landfill leachate. *Water Research* 159: 38–45. https://doi.org/10.1016/J.WATRES.2019.04.060.

Hidalgo-Ruz, V., Gutow, L., Thompson, R.C. and Thiel, M. 2012. Microplastics in the marine environment: A review of the methods used for identification and quantification. *Environmental Science and Technology* 46(6): 3060–3075. https://doi.org/10.1021/ES2031505.

Hwang, B.B. 2007. Unpacking the Packaging Problem: an international solution for environmental impact of packaging waste. Unpublished paper. University of Baltimore.

Huang, J. and Boui, H. 2001. Structural changes in the demand for food in Asia: empirical evidence from Taiwan. *Agricultural Economic* 26(1): 57–69.

ICN. 2007. http://www.datacon.co.id/Plastic%20Resin.html

Imhoff, D. 2005. Paper or plastics: searching for solutions to an over packed worldhttp://www.watershedmedia.org/pop_abstract.html.

INCPEN, the Industry Council for Packaging and the Environment, Responsible Packaging, code of practice to optimizing packaging and minimizing waste, 2003. Retrieved at http://www.incpen.org/pages/data/CodeofPractice.pdf.

Isobe, A., Uchiyama-Matsumoto, K., Uchida, K. and Tokai, T. 2017. Microplastics in the Southern Ocean. *Marine Pollution Bulletin* 114(1): 623–626. https://doi.org/10.1016/J.MARPOLBUL.2016.09.037.

Jambeck, J.R., Geyer, R., Wilcox, C., Siegler, T.R., Perryman, M., Andrady, A., Narayan, R. and Law, K.L. 2015. Plastic waste inputs from land into the ocean. *Science* 347(6223): 768–771. https://doi.org/10.1126/science.1260352.

Jambeck, J.R., Ji, Q., Zhang, Y.-G., Liu, D., Grossnickle, D.M. and Luo, Z.-X. 2015. Plastic waste inputs from land into the ocean. *Science* 347(6223): 764–768. http://www.sciencemag.org/cgi/doi/10.1126/science.1260879.

Ji, Z.-G. 2008. *Hydrodynamics and Water Quality*. John Wiley and Sons, Inc. https://doi.org/10.1002/9780470241066.

Kawecki, D. and Nowack, B. 2020. A proxy-based approach to predict spatially resolved emissions of macro- and microplastic to the environment. *Science of The Total Environment* 748: 141137. https://doi.org/10.1016/J.SCITOTENV.2020.141137.

Khatmullina, L. and Isachenko, I. 2017. Settling velocity of microplastic particles of regular shapes. *Marine Pollution Bulletin* 114(2): 871–880. https://doi.org/10.1016/j.marpolbul.2016.11.024.

KepmenLHK P.75/MENLHK/SETJEN/KUM.1/10/2019 . Peta Jalan Pengurangan Sampah oleh Produsen. Peraturan Menteri Lingkungan Hidup dan Kehutanan.

Klein, S., Worch, E. and Knepper, T.P. 2015. Microplastics in the Rhine-Main area in Germany: Occurrence , spatial distribution and sorption of organic contaminants. *Environmental Science and Technology* 49(0): 2–3.

Kooi M, Besseling E., Kroeze, C., van Wezel, A.P. and K.A. 2017. Modeling the Fate and Transport of Plastic Debris in Freshwaters: Review and Guidance. *Springer, Heidelberg* 125–152. https://doi.org/10.1007/978-3-319-61615-5_14.

Lebreton, L. and Andrady, A. 2019. Future scenarios of global plastic waste generation and disposal. *Palgrave Communications* 5(1). https://doi.org/10.1057/s41599-018-0212-7.
Lebreton, L.C.M., van der Zwet, J., Damsteeg, J.W., Slat, B., Andrady, A. and Reisser, J. 2017. River plastic emissions to the world's oceans. *Nature Communications* 8. https://doi.org/10.1038/ncomms15611.
Lu, H.C., Ziajahromi, S., Neale, P.A. and Leusch, F.D.L. 2021. A systematic review of freshwater microplastics in water and sediments: Recommendations for harmonisation to enhance future study comparisons. In *Science of the Total Environment* (Vol. 781). Elsevier B.V. https://doi.org/10.1016/j.scitotenv.2021.146693.
Lusher, A.L., Tirelli, V., O'Connor, I. and Officer, R. 2015. Microplastics in Arctic polar waters: The first reported values of particles in surface and sub-surface samples. *Scientific Reports* 5. https://doi.org/10.1038/srep14947.
Masry, M., Rossignol, S., Gardette, J., Therias, S., Bussière, P. and Wah-Chung, P.W. 2021. Characteristics, fate, and impact of marine plastic debris exposed to sunlight: A review, Marine Pollution Bulletin 171, 112701. https://doi.org/10.1016/j.marpolbul.2021.112701
McGlade, J.2004. European Packaging Waste Trends and the Role of Economic Instruments.https://www.eea.europa.eu/media/speeches/01-03-2004
Napper, I.E., Bakir, A., Rowland, S.J. and Thompson, R.C. 2015. Characterisation, quantity and sorptive properties of microplastics extracted from cosmetics. *Marine Pollution Bulletin* 99(1–2): 178–185. https://doi.org/10.1016/J.MARPOLBUL.2015.07.029.
Ng, K.L., Obbard, J., Privalence of Microplastoics in Singapore Coastal Marine Environment, Marine Pollution Bulletin 52(7): 761–7. https://doi.org/10.1016/j.marpolbul.2005.11.017
Nizzetto, L., Bussi, G., Futter, M.N., Butterfield, D. and Whitehead, P.G. 2016. A theoretical assessment of microplastic transport in river catchments and their retention by soils and river sediments. *Environmental Science: Processes and Impacts* 18(8): 1050–1059. https://doi.org/10.1039/c6em00206d.
Quik, J.T.K., de Klein, J.J.M. and Koelmans, A.A. 2015. Spatially explicit fate modelling of nanomaterials in natural waters. *Water Research* 80: 200–208. https://doi.org/10.1016/j.watres.2015.05.025.
Ramadan, A.H. and Sembiring, E. 2020. Occurrence of Microplastic in surface water of Jatiluhur Reservoir. *E3S Web of Conferences* 148: 1–4. https://doi.org/10.1051/e3sconf/202014807004.
Reuters. 2009. China's plastic bag ban kicks in to mixed response. http://www.reuters.com/article/latestCrisis/idUSPEK210811. Accessed 13 October 2009.
Ritchie, H. 2018. How much plastics and waste do we produce? Ourworldindata.org.https://ourworldindata.org/faq-on-plastics#how-much-plastic-and-waste-do-we-produce.
Rochman, C.M. 2013. Plastics and priority pollutants: A multiple stressor in aquatic habitats. *Environmental Science and Technology* 47(6): 2439–2440. https://doi.org/10.1021/es400748b.
Rochman, C.M., Hoh, E., Kurobe, T. and Teh, S.J. 2013. Ingested plastic transfers hazardous chemicals to fish and induces hepatic stress. *Scientific Reports* 3: 1–7. https://doi.org/10.1038/srep03263.
Schwarz, A.E., Ligthart, T.N., Boukris, E. and van Harmelen, T. 2019. Sources, transport, and accumulation of different types of plastic litter in aquatic environments: A review study. *Marine Pollution Bulletin* 143: 92–100. https://doi.org/10.1016/J.MARPOLBUL.2019.04.029
Sembiring, E. and Fareza, A.A., Suendo, V. and Reza, M. 2020. The Presence of Microplastics in Water, Sediment, and Milkfish (Chanoschanos) at the Downstream Area of Citarum River, Indonesia. *Water, Air, and Soil Pollution* 231: 7. https://doi.org/10.1007/s11270-020-04710-y.
Song, Y.K., Hong, S.H., Jang, M., Han, G.M., Jung, S.W. and Shim, W.J. 2017. Combined Effects of UV Exposure Duration and Mechanical Abrasion on Microplastic Fragmentation by Polymer Type. *Environmental Science and Technology* 51(8): 4368–4376. https://doi.org/10.1021/acs.est.6b06155.
Ta, A.T. and Babel, S. 2020. Microplastics pollution with heavy metals in the aquaculture zone of the Chao Phraya River Estuary, Thailand. *Marine Pollution Bulletin* 161 (October): 111747. https://doi.org/10.1016/j.marpolbul.2020.111747.
The Vietnamese plastic industry. http://74.125.153.132/search?q=cache:cwDxC8kcQVwJ:www.ambhanoi.um.dk/NR/rdonlyres/20C16B40-53AD40139F66503944D9706B/0/TheVietnamesePlasticsIndusty.doc+afpi+plasticsandcd=10andhl=jaandct=clnkandgl=jp.
Thompson, R.C., Moore, C.J., Saal, F.S.V. and Swan, S.H. 2009. Plastics, the environment and human health: Current consensus and future trends. *Philosophical Transactions of the Royal Society B: Biological Sciences* 364(1526): 2153–2166. https://doi.org/10.1098/RSTB.2009.0053.

van Cauwenberghe, L., Vanreusel, A., Mees, J. and Janssen, C.R. 2013. Microplastic pollution in deep-sea sediments. *Environmental Pollution* 182: 495–499.

van Emmerik, T., Kieu-Le, T.C., Loozen, M., Oeveren, K. van, Strady, E., Bui, X.T., Egger, M., Gasperi, J., Lebreton, L., Nguyen, P.D., Schwarz, A., Slat, B. and Tassin, B. 2018. A methodology to characterize riverine macroplastic emission into the ocean. *Frontiers in Marine Science* 5(OCT): 1–11. https://doi.org/10.3389/fmars.2018.00372.

van Sebille, E., Griffies, S.M., Abernathey, R., Adams, T.P., Berloff, P., Biastoch, A., Blanke, B., Chassignet, E.P., Cheng, Y., Cotter, C.J., Deleersnijder, E., Döös, K., Drake, H.F., Drijfhout, S., Gary, S.F., Heemink, A.W., Kjellsson, J., Koszalka, I.M., Lange, M., … Zika, J.D. 2018. Lagrangian ocean analysis: Fundamentals and practices. *Ocean Modelling* 121(October 2017): 49–75. https://doi.org/10.1016/j.ocemod.2017.11.008.

van Wezel, A., Caris, I. and Kools, S.A.E. 2016. Release of primary microplastics from consumer products to wastewater in the Netherlands. *Environmental Toxicology and Chemistry* 35(7): 1627–1631. https://doi.org/10.1002/etc.3316.

Wang, G., Lu, J., Li, W., Ning, J., Zhou, L., Tong, Y., Liu, Z., Zhou, H. and Xiayihazi, N. 2021. Seasonal variation and risk assessment of microplastics in surface water of the Manas River Basin, China. *Ecotoxicology and Environmental Safety* 208: 111477. https://doi.org/10.1016/j.ecoenv.2020.111477.

Waste Watch, Waste on line, 2006. Retrieved at http://www.wasteonline.org.uk/resources/InformationSheets/Packaging.htm.

Wrap that in plastic? Not in Taiwan, unless you pay. http://www.csmonitor.com/2004/0615/p07s02-woap.html.

Xanthos, D. and Walker, T.R. 2017. International policies to reduce plastic marine pollution from single-use plastics (plastic bags and microbeads): A review. *Marine Pollution Bulletin* 118(1–2): 17–26. https://doi.org/10.1016/J.MARPOLBUL.2017.02.048.

Yadav, V., Sherly, M.A., Ranjan, P., Tinoco, R.O., Boldrin, A., Damgaard, A. and Laurent, A. 2020. Framework for quantifying environmental losses of plastics from landfills. *Resources, Conservation and Recycling* 161: 104914. https://doi.org/10.1016/J.RESCONREC.2020.104914.

Zhixin, Z., Keng, H., Xiaoua, Q., Xianyu, L., Xiandong, Z. et al. 2002. China Statistical Year Book. China Statistics Press. http://www.stats.gov.cn/english/statisticaldata/yearlydata/YB2002e/ml/inde

Zhu, K., Jia, H., Sun, Y., Dai, Y., Zhang, C., Guo, X., Wang, T. and Zhu, L. 2020. Long-term photo-transformation of microplastics under simulated sunlight irradiation in aquatic environments: Roles of reactive oxygen species. *Water Research* 173: 115564. https://doi.org/10.1016/J.WATRES.2020.115564.

Chapter 3

The Correlation Between Plastic and Climate Change

Negin Ficzkowski and *Gail Krantzberg**

INTRODUCTION

While biodiversity and ecosystem degradation are often the focus of plastic-related research, the contribution of plastic to the cumulative climate-relevant trace gases is often unrecognized. As such, the plastic crisis is much less commonly viewed as a contributor to climate change. The compounding impact of plastic production and waste is only expected to increase as more plastic is produced and accumulated in the environment. The problem of conventional plastic feedstocks is the core issue of the linear economy and its associated climate change stress ors. This chapter primarily explores the lifecycle impacts of plastics on cumulative greenhouse gas emissions. It then investigates the cyclical nature of plastic pollution and extreme weather events. The chapter also illustrates, in brief, the impacts of plastic pollutants on the largest carbon sinks on earth and investigates the adverse effects on global food supplies. Finally, it discusses landfill emission contributions and other land-related complications as they relate to plastic.

Decades of observations, research, and scientific learning about climate change have proven the significant risks that climate change holds for people, ecosystems, and natural resources (National Research Council, 2010). Current concentrations of greenhouse gases are the result of progressive ecosystem degradation, acidification of oceans, and species extinctions (Webb, 2012). Increases in atmospheric carbon have been associated with temperature rises and abrupt changes in climate, land, and water resources (Webb, 2012). Climate change also interacts in complex ways with other ongoing changes within ecological ecosystems, such as land use, agriculture, and food production, which are all bidirectionally interconnected with climate change (National Research Council, 2010). Similarly, there is growing evidence proving that the global plastic crisis and climate change exacerbate one another (Ford et al.,

McMaster University, 1280 Main St. W., Hamilton ON L8S 4K1.
* Corresponding author: krantz@mcmaster.ca

2021). The inherent link between the climate crisis and plastics in society should lead scientists and policymakers to tackle them in unison and use a holistic approach as opposed to viewing them as distinct issues (Ford et al., 2021).

Often plastics become waste for landfills after a comparatively brief usage in their lifetime, and macroplastics, which are conventional plastic waste larger than 25 mm, find their way to surface waters, eventually accumulating in aquatic systems intact for a long time or breaking down into microplastics (MPs) over time (van Reenen, 2020). Ultraviolet (UV) radiation plays a key role in plastic fragmentation. Because UV light is absorbed rapidly by water, plastics generally take much longer to degrade at sea than on land (Andrady, 2015).

Absorbing around 23 ± 5% of annual greenhouse gas emissions generated by human activities over the 2009–2018 decade and more than 90% of the excess heat in the climate system, oceans are one of the most essential ecosystems in mitigating the impacts of climate change (United Nations 2020). Earth's oceans have already absorbed 20% to 40% of all anthropogenic carbon emitted since the dawn of the industrial era (Webb, 2012). Phytoplankton and zooplankton play a critical role in the biological carbon pump that captures carbon at the ocean's surface and transports it into the deep oceans (Fig. 1), preventing it from re-entering the atmosphere (Hamilton et al., 2019). Furthermore, phytoplankton has a critical and foundational role as the first link in the global food web of all species, feeding other species like zooplankton, which then feed much larger species like fish, sea birds, and whales. Today, we know that not only are MPs outnumbering the zooplanktons in some areas but also some of the additives used to modify the properties of plastics are biologically active, destroying aquatic systems, and affecting the development and reproduction in many species (Oehlmann et al., 2009; Meeker et al., 2009).

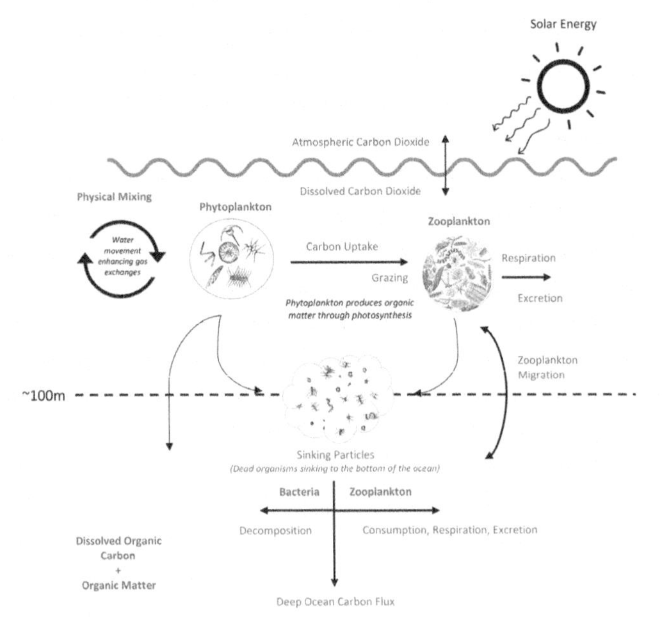

Fig. 1. Simplified Mechanism of Ocean Biological and Physical Carbon Pump. Image Modified From Adaptation of a New Wave of Ocean Science (NASA) and Ocean & Climate Platform.

In addition to endangering aquatic biodiversity, plastic pollution impacts the livelihood of those relying on marine resources and causes a range of health issues for those consuming seafood infested with toxic micro- and nanoplastics (World Health Organization, 2019). Women, in particular, suffer from plastic-related toxicity risk due to higher aggregate exposure to plastics in various ways, including feminine care and hygiene products which places them at a high risk of miscarriages and cancer (UNEP, 2021). The analysis of the MPs also reveals pollution from fabrics and polyester clothing which breaks off and enters the water supply after each wash cycle (Hou et al., 2021); it is difficult to see, yet it shows the destructive impact of plastic.

Despite numerous well-published studies about the destructive terrestrial and marine effects of plastic, nearly 367 million metric tons of conventional plastic are produced per annum globally (Fig. 2). Every year, an estimated 5 to 12 million metric tons of plastic enters the oceans, 89% of which are single-use items (UNEP, 2020). Up to 80% of the plastics found in marine systems originate from rivers that transport plastics from land to ocean (Kosior and Crescenzi, 2020). Recent studies suggest that the levels of plastic pollution in freshwater sources are comparable to those found in oceans (Dris et al., 2015; Earn et al., 2021).

As Fig. 3 illustrates, less than only 10% of all plastic waste is recycled annually, while on average 8 million tons are being disposed of into landfills (van Reenen, 2020). While better than landfilling, recycling can be hardly viewed as the golden standard of combating plastic waste as it introduces relatively lower quality polymer-based materials back into the system which not only breaks down into MPs easier than the virgin plastics but is also not cost- and energy-comparative with the production of new, virgin plastic (Tolinski, 2012).

This chapter offers a critical evaluation of the bidirectional interlink between plastic crises and climate change to inform global and domestic policy for devising a

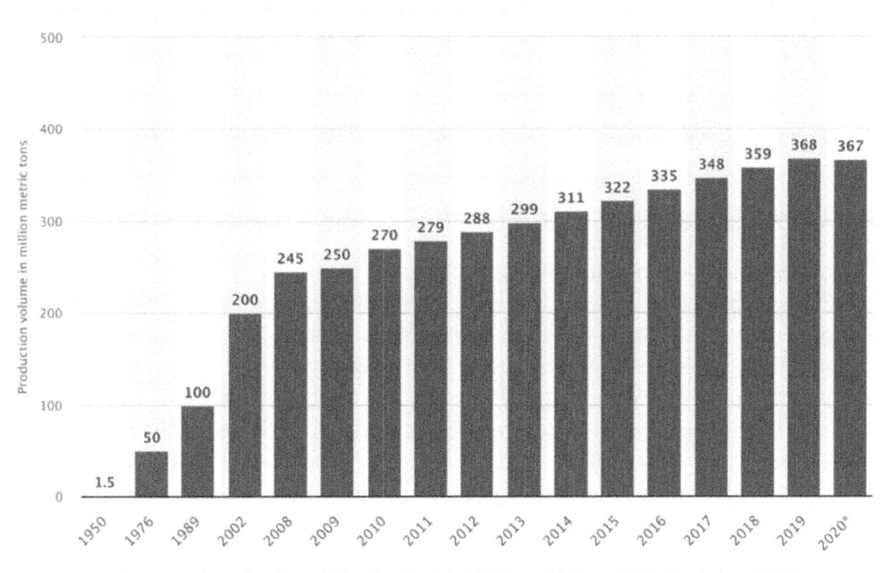

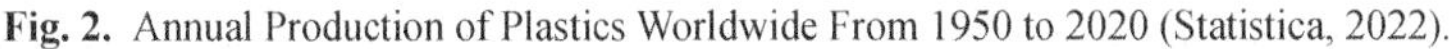

Fig. 2. Annual Production of Plastics Worldwide From 1950 to 2020 (Statistica, 2022).

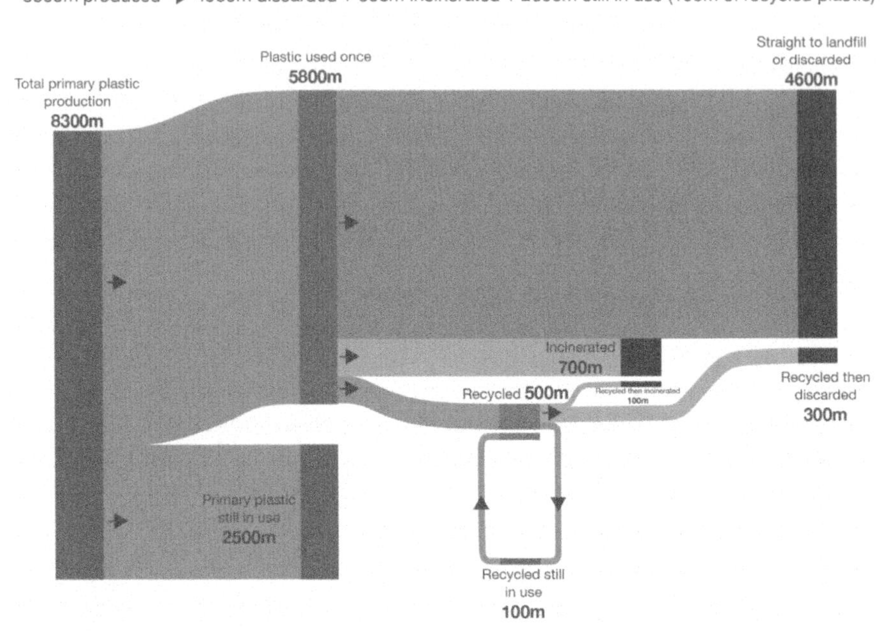

Fig. 3. Global Plastic Production and Its Fate (1950–2015). Source: Our World in Data.

joint solution for addressing both problems (Ford et al., 2021). The interconnection is discussed in five major areas based on the threats of climate change which are as follows:

1. Cumulative greenhouse gas emissions of the plastic lifecycle
2. The impacts of extreme weather events associated with the changing climate
3. Ocean acidification, pollution, and biodiversity loss
4. Health impacts and ecosystem vulnerability to the effects of climate change as well as plastic pollution
5. Land-use implications

Cumulative Greenhouse Gas Emissions and Carbon Budgets

Increasing temperatures and GHG concentrations are the main drivers of the multitude of changes observed in the earth system, including decreases in the amounts of ice stored in mountain glaciers and polar regions, increases in sea level, changes in ocean chemistry, and changes in the frequency and intensity of heat waves, precipitation events, and droughts (Webb, 2012). Projections of future climate change indicate that earth will continue to warm unless we not only significantly limit the activities that release heat-trapping greenhouse gases into the atmosphere but also retroactively mitigate the current state through carbon sequestration. Over the last several decades, the chemical signature of the excess carbon dioxide (CO_2) in the atmosphere has been linked to the composition of the CO2 in emissions from the burning of fossil fuels (coal, oil, and natural gas), marking it the single largest human driver of climate change (Webb, 2012).

Plastics play a key role in the petrochemical and oil industries since the primary feedstock for conventional plastic production is fossil fuel derivates (Hamilton et al., 2019). According to International Energy Agency (IEA), plastics (thermoplastics and textiles combined) can make up to two-thirds of the demand for oil in the sector and the entire demand growth for oil by 2015 (IEA, 2018). According to BP (2019), on average between 2015 and 2020, the total amount of oil used in the petrochemical sector on average was 15 million barrels per day (mbpd), the total oil demand for plastics was around 10 mbpd (~67%), and it will rise over time to 77% in 2040. This fact positions plastic production as a major threat to future climate change projections. Most emission reports highlight five different contributing stages to the plastic lifecycle: a) fossil fuel extraction and transport (also known as cradle-to-resin); b) refinement and manufacturing (both resin manufacturing and plastic product manufacturing); c) usage and consumption; d) end-of-life management; and e) after-life emissions or the ongoing impacts of plastic in the environment as it degrades (Ford et al., 2021).

In 2015, the primary production of plastic emitted the equivalent of more than a billion metric tons of carbon dioxide (Geyer, 2020).[1] It is estimated that by 2050, the greenhouse gas emissions from plastic could reach over 56 gigatons, accounting for 10 to 13% of the entire remaining carbon budget globally (Hamilton et al., 2019). Limitations in the availability and accuracy of by-sector data make it challenging to obtain accurate emission amounts per lifecycle stage per sector; however, it is conservatively estimated that amongst all stages, extraction and refining are the most

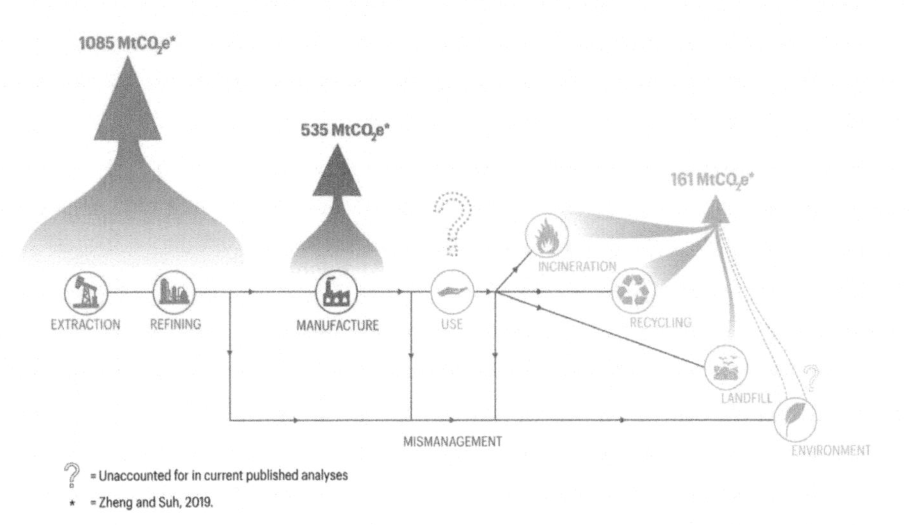

Fig. 4. GHG Emission Profile for Plastic Lifecycle (Ford et al., 2021).

[1] Geyer takes the most comprehensive definition of plastics in 2017 at 438 mt and notes that 93% of finished plastics are from polymers (almost all from feedstocks defined as oil) and 7% from additives.

GHG-expensive stages, followed by end-of-life management (Hamilton et al., 2019). Figure 4 illustrates this distribution. The emissions from the use phase vary based on the functional unit, volume, and weight as well as the functions of the plastic product, and they are not typically accounted for in high-level analyses (Ford et al., 2021). Further, indirect emissions or potential savings during the lifecycle may be considered where GHG savings are enabled due to the lightweight properties of plastic that result in lower emissions during transport which are relative to other materials, such as glass, wooden, or metal items (Ford et al., 2021).

From a carbon budget standpoint, according to the Ocean Conservancy, high-income countries, including those in Europe and North America, have been exporting their plastic waste to the top plastic emitting countries (namely China, Indonesia, the Philippines, Thailand, and Vietnam) for recycling and to reduce their annual emissions from waste handling processes (Fig. 5). In these countries, much of the soft plastic is burned illegally as a cheap fuel source of energy or sold to cement plants for the same purpose; any material that is of little use to businesses will be lost from the underdeveloped waste management system by entering the ocean (Ocean Conservancy, 2015).

Efforts to quantify emissions from plastic degradation are still in the early stages, but they already demonstrate the continual release of methane and other greenhouse gases at the ocean's surface, and the increase of these emissions as the plastic breaks down further (Hamilton et al., 2019). Further, Royer et al. (2018) have demonstrated that plastic on coastlines, riverbanks, and landscapes releases greenhouse gases.

Biodiversity Loss, Ocean Acidification, and Impacts on the Largest Carbon Sink on Earth

Plastic pollution is a significant driver of marine biodiversity loss and so are the impacts of climate change (Fig. 6). The reefs and other vulnerable habitats are suffering from ocean warming and acidification (Ford et al., 2021), while the ecosystem is continuously exposed to an influx of harmful plastic pollutants. By 2012, scientists have documented various harmful impacts of marine plastic pollution on 800 different marine species, including all known species of sea turtles, half of all species of marine mammals, and one-fifth of all species of seabirds (Secretariat of the Convention on Biological Diversity, 2012). Globally, at least 23% of marine mammal species, 36% of seabird species, and 86% of sea turtle species are known to be affected by plastic debris (Stamper et al., 2009). In addition, MP in the oceans can contaminate the phytoplankton community and interfere with the oceans' capacity to sequester carbon dioxide through photosynthesis and nutrient cycling (Royer et al., 2018). Laboratory experiments suggest that plastic pollution can also reduce the metabolic rates, reproductive success, and survival of zooplankton that transfer the carbon to the deep ocean (Hamilton et al., 2019). Zooplankton ingestion of MPs can also result in more algal growth leading to more organic particle remineralization and a decrease in water column oxygen inventory loss (Kvale et al., 2021).

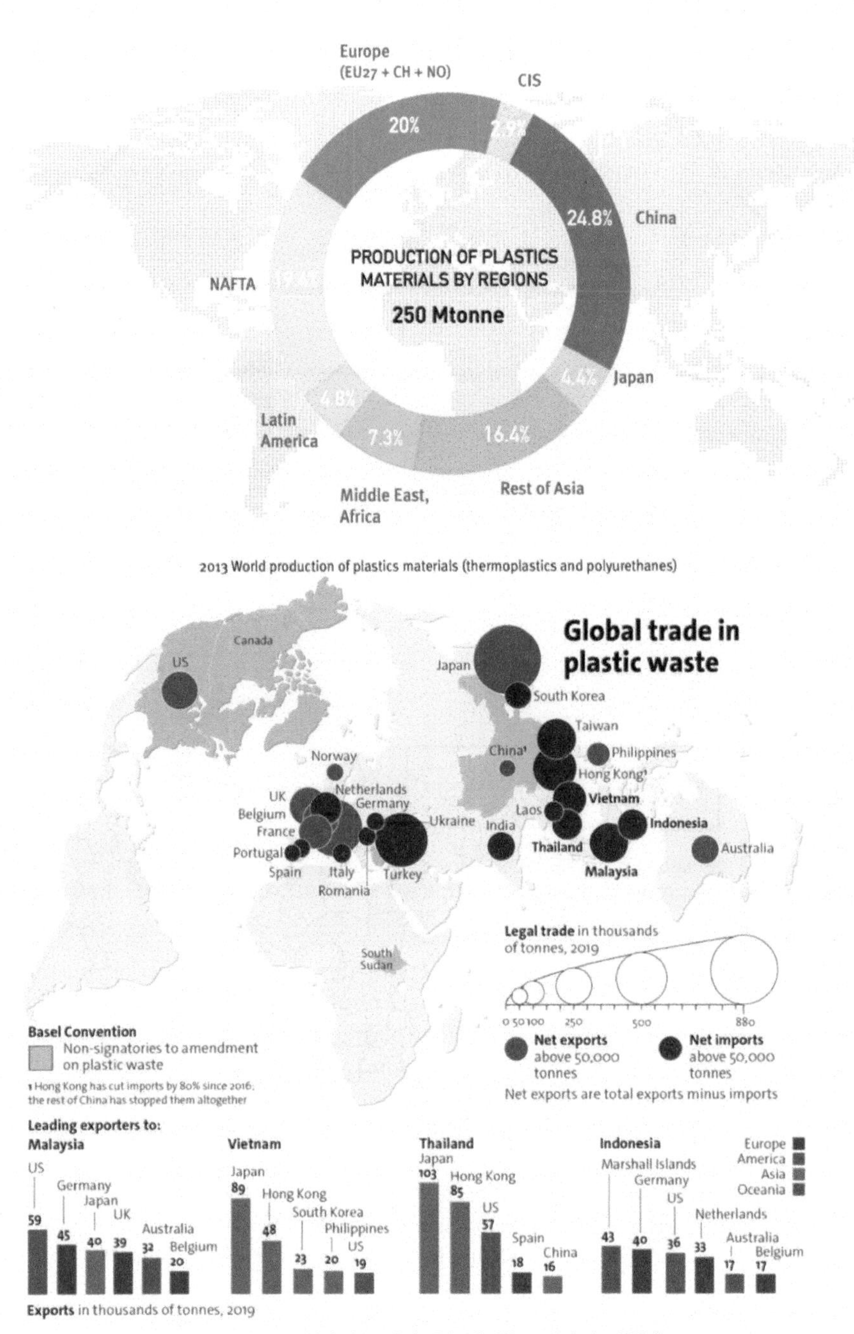

Fig. 5. Global Trade in Plastic Waste (Marin, 2021).

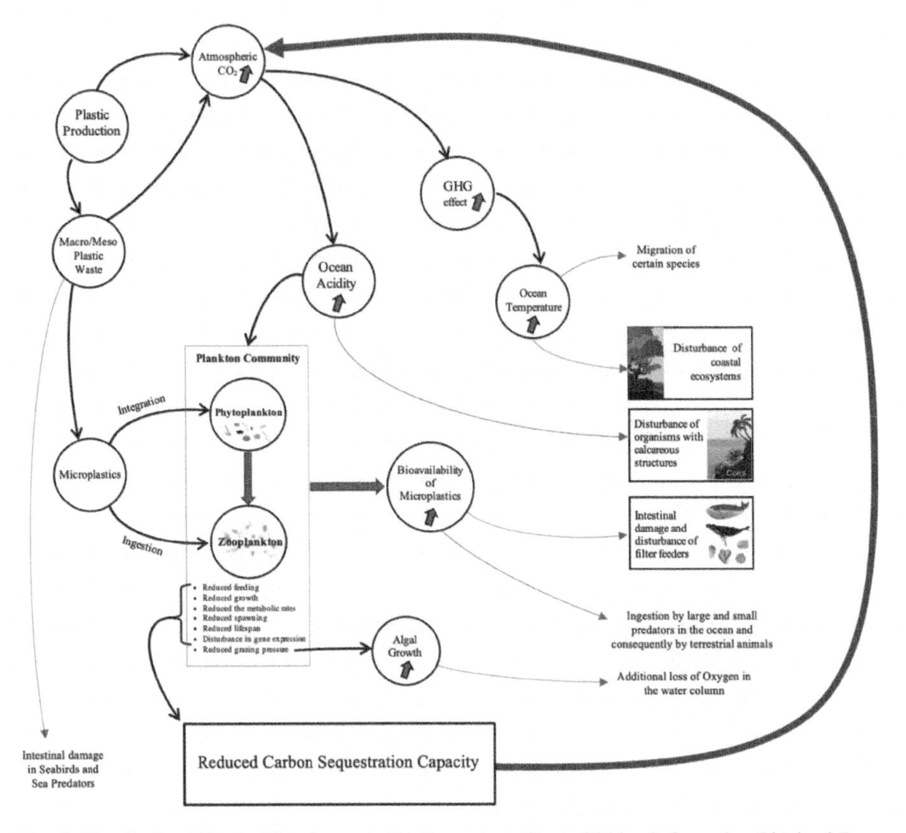

Fig. 6. The Cycle of Plastic Disturbance in the Ecosystem (Created Using Information Obtained From (Ocean and Climate Platform, n.d.; Ford et al., 2021; Royer et al., 2018; Kvale et al., 2021; Kosior and Crescenzi, 2020).

MPs have also been identified as an emerging threat to much larger organisms, like whales that are exposed to MP ingestion as a result of their filter-feeding activity (Kosior and Crescenzi, 2020). Macro-plastics (plastic objects larger than 25 mm) are just as big of a problem as micro- (less than 5 mm) and meso-plastics (between 5 to 25 mm). In many locations, dead animals are found with stomachs full of plastic, choked by packaging remains, entanglement, or starved from reduced appetite resulting from plastic ingestion (Fig. 7). Other harmful effects are blockage of the digestive tract and internal injuries that are particularly evident in sea turtles that tend to eat plastic bags mistaking them for jellyfish (Kosior and Crescenzi, 2020). Records of marine turtles ingesting plastic bags date back to the late 1950s followed by a series of discoveries and reported evidence between 1960 to 1974 (Ryan, 2015). During the 1970s, various instances of plastic fragments were reported in marine habitats, resulting in the first scientific journal publication on wildlife entanglement (Thompson et al., 2009). In 1984, the Honolulu Strategy was introduced as a framework document to internationally reduce plastic waste and devise a management structure to meet the specific needs of different countries including Canada (Ryan, 2015).

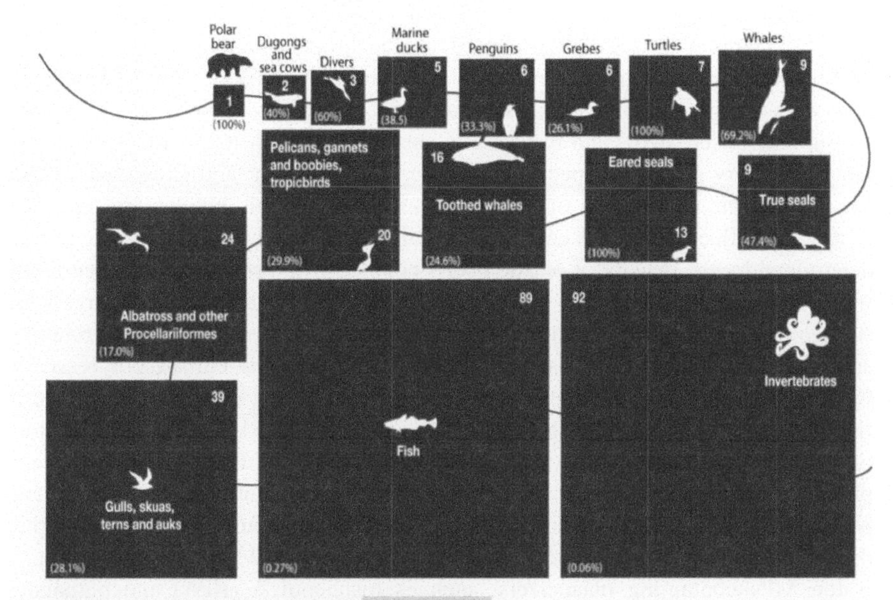

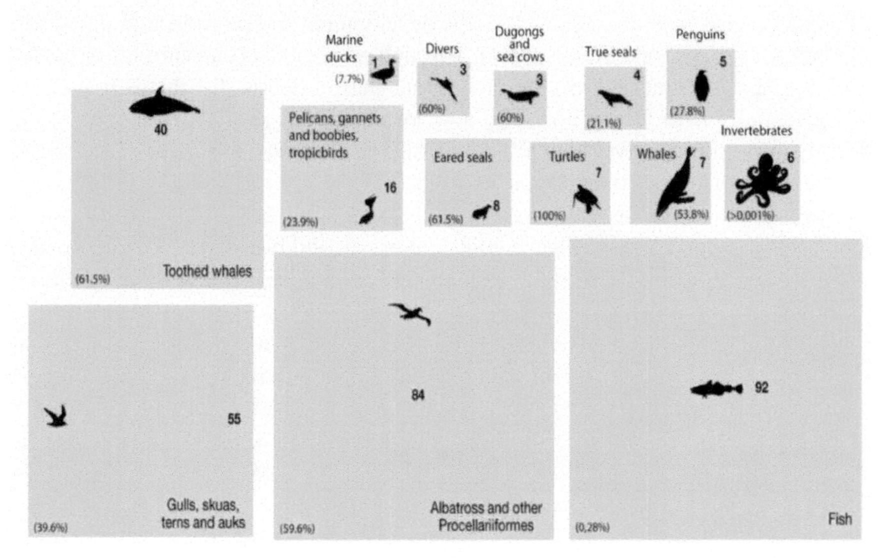

Fig. 7. Number of species with documented records of entanglement in and ingestion of marine debris, figure created by Maphoto/Riccardo Pravettoni based on data from Kuhn et al. 2015, obtained from GRID-Arendal.

Adverse Effects of Plastic on Global Food Supplies, Global Health, and Economy

The impact of plastic on the nutrient cycle is not limited to the plankton community in the ocean (Jâms et al., 2020). Plastic particles can attract insoluble organic chemicals in the ocean (including many known toxic substances) (Kuhn et al., 2015), introducing persistent organic pollutants (POPs) into marine foodwebs (Cole et al., 2011; Ivar do Sul and Costa, 2014) and resulting in subsequent health issues for humans and other predatory terrestrial species consuming them (Kosior and Crescenzi, 2020). The geographic distribution of MP debris available for this involuntary ingestion is strongly influenced by the entry points and transport pathways, which are in turn determined by the density of plastic debris coupled with prevailing ocean currents, wind, and waves (van Sebille et al., 2015). Ocean currents are created largely by surface winds and partly by temperature and salinity gradients, the earth's rotation, and tides. Gradual sea level rise resulting from the changing climate progressively worsens the impacts of high tides, surges, and waves resulting from storms, freshwater floods, and coastal mountain catchments (Cayan et al., 2008).

Due to their hydrophobic nature, MPs absorb polybrominated diphenyl esters, pharmaceuticals and personal care products with polychlorinated biphenyls, and polycyclic aromatic hydrocarbons with concentrations that are 10^5–10^6 times higher than in the surrounding water column (Krantzberg, 2020). When consumed through water, MPs containing plasticizers, such as bisphenol A (BPA) and phthalates (commonly found in plastics ranging from food storage containers and metal can liners to baby bottles and sippy cups), have been found to be endocrine-disrupting and therefore affect the sexual development of children born to exposed mothers, the hatching success in animals as well as the development and reproduction in wildlife and humans (Klaper and Welch, 2011). Through waves and ocean currents or having been ingested by marine biota in one region, plastic debris distributes non-native and potentially harmful organisms to new locations (Kosior and Crescenzi, 2020), directly impacting the livelihood of communities reliant on fisheries and haunting as well as the economy of their surrounding regions.

In addition to the health and traditional economy of these regions, the floating plastic debris interferes with the chemical, physical, and biological processes within blue carbon ecosystems comprising tidal marshes, mangrove forests, kelp forests, seagrass meadows, and coastal wetlands and downgrading their capacity for global climate change adaptation and mitigation goals for at least 28 countries globally (Adyel and Marcreadie, 2022) as well as interferes with climate resilience in these communities (Fig. 8). Conserving, protecting, and restoring Blue Carbon ecosystems should not only be an integral part of climate action plans at all levels of governance globally (Hilmi et al., 2021) but should also be interjoined with global action to stop the release of plastics into the environment (Fig. 9).

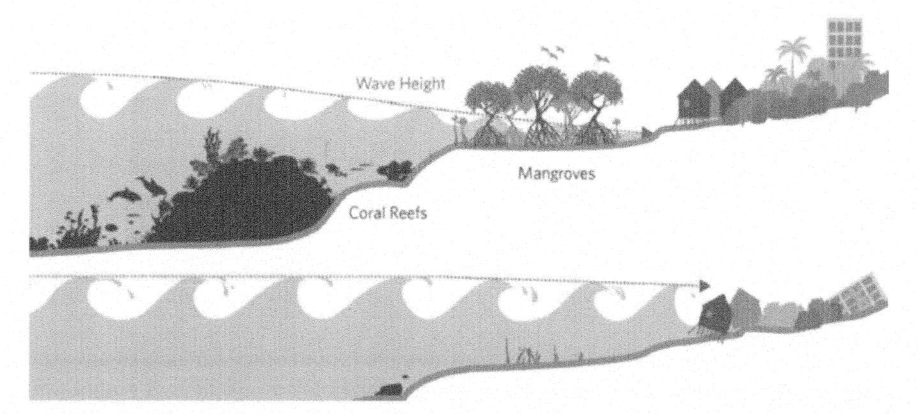

Fig. 8. Illustration of Climate Resilience Through Blue Carbon Ecosystems. Image Acquired From the Policy Brief for Urging G20 to Strengthen the Role of Blue Carbon in Climate Action (Mansouri et al., 2020).

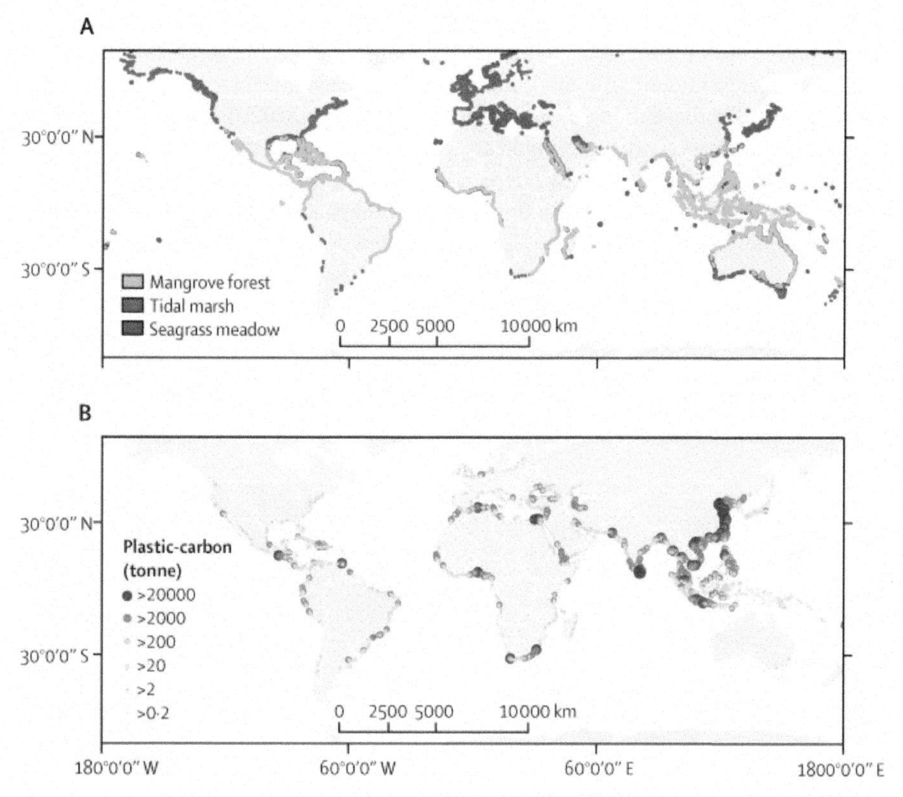

Fig. 9. Global Distribution of Blue Carbon Ecosystems (Top) Compared With Mass of Plastic-Carbon Discharged to Global Oceans Through Rivers and Coastal Areas (Bottom) (Adyel and Marcreadie, 2022).

Extreme Weather Events, Warmer Temperatures, Droughts, and Storms

As functionally designed for lightweight properties, loose plastic parts can be carried and moved around quite easily. Even during a regular windy waste collection day in a Canadian municipality, household plastic waste can be spotted dispersed down the street (Fig. 10). Extreme weather events associated with climate change (including flooding, tropical storms, stronger winds, more frequent rain events, and sea level rise) influence the distribution of mismanaged plastic waste and MP pollution into remote areas and between terrestrial, freshwater, and marine environments (Ford et al., 2021).

Various examples in the literature illustrate the bidirectional impacts of extreme weather events and mismanaged plastic waste globally. For instance, a typhoon in Sanggou Bay, China, increased the abundance of MPs within seawater and sediments by as much as 40% (Wang et al., 2019). Similarly, increased rainfall, associated with monsoons, is estimated to increase the MP concentration entering the Bay of Bengal from the Ganges at approximately 1 billion MPs per day during the pre-monsoon season and 3 billion post-monsoon seasons (Napper et al., 2021). Further inputs of terrestrial plastic into aquatic environments may release plastics trapped in coastal sediments and increase the risk of flooding (Sebille et al., 2020). The floating waste can clog the drainage system during heavy rain and cause flooding and consequential damages (Galgani et al., 2015; Welden and Lusher, 2017).

There are potentially infinite loops created by the linkage between climate change and plastic pollution. For instance, it has been shown that lower perceptions of the quality of tap water increase the probability of purchasing bottled water which is one of the leading sources of plastic pollution (Zapata, 2021). The same study illustrates that the increase in temperature due to global warming is associated on average with an increase of almost one-fifth of a water bottle (Zapata, 2021). Another aspect is the insufficient infrastructure and proper water supply system in rural areas, creating a disorganized management network that fails to meet the demand for drinking water

Fig. 10. Illustration of plastic waste dispersed by wind in Halton region (left) and city of Hamilton (right).

and consequently increasing dependency on water transportation methods using plastic bottles (Zapata, 2021). Lacking a solid waste management system, putting this and similar plastic waste into the system blocking drainage, and exacerbating flooding during heavy rainfall. Waste wash-off into roads and waterways, leachate and polluted waters requiring excess processes, sewage overflow, and significant demolish waste generated from damages are only a few foreseeable impacts of water disasters and sea-level rise within coastal areas.

Plastics do not only end up in the sea but on the beaches of some of the remotest, uninhabited islands in the Pacific, such as Henderson Island in the Pitcairn group (Kosior and Crescenzi, 2020). In the Cilician Basin located in the Northeastern Mediterranean Sea, plastic items comprise more than 80% of the dominant material type in the coastal environments up to 4.23% of which was transboundary litter transported with currents from neighboring countries (Aydın et al., 2016).

In 2015, scientists at the University of Utrecht in the Netherlands used sparse data from drifting buoys to estimate the amount of plastic at the surface ocean and compared this to World Bank estimates. Nearly 99% of the plastic is no longer on the surface of the ocean, rather it is found in the icy mountain peaks, at the bottom of the ocean, inside various marine species, or broken down into MPs (van Sebille et al., 2015). The melting of sea ice due to climate change has been linked to an increase in plastic pollution, as the ice works as a trap for MPs that when melted release them all into the earth's oceans (Ford et al., 2021).

Land Management Interference

According to the 2019 special report from the Intergovernmental Panel on Climate Change (IPCC), the land is critically important in the discussions of climate action, both as a source of greenhouse gas emissions and as a solution for climate change (IPCC, 2019). About 23% of global human-caused greenhouse gas emissions come from agriculture, forestry, and other land uses. Additionally, 44% of recent human-driven methane, a potent greenhouse gas, came from agriculture, peatland destruction and other land-based sources (IPCC, 2019). That said, well-managed land-based emissions-reduction efforts, such as afforestation, reforestation, or wetland rehabilitation, have great potential in carbon sequestration and other co-benefits for the ecosystem in face of the climate change (IPCC, 2019), provided that the continuous flow of plastic waste generation does not overwhelm GHG abatement measures.

The plastic crisis like many other destructive human activities starts on land, leaking into soil and water bodies following a series of inefficient and underdeveloped end-of-life management activities (Kosior and Crescenzi, 2020). Before 1988, plastic waste was partially disposed of directly into the ocean leading to enforced restrictions on dumping of garbage from ships by the International Convention for the Prevention of Pollution from Ships (MARPOL 73/78) as of December 31, 1988. Over the last two decades, according to a 2015 global study, less than 20% of the annual 8 million tons of marine plastics come from ocean-based sources, like fisheries and fishing vessels (Jambeck et al., 2015); the remaining 80% originates from land-based sources, as illustrated in Fig. 11 (GESAMP, 2015). According to the Ocean

Fig. 11. Plastic Waste Inputs From Land to the Oceans (Jambeck et al., 2015).

Conservancy (2015), 75% of ocean plastic debris that originates from land-based sources comes from uncollected waste directly deposited into and around rivers and other water bodies that function as direct pathways into marine ecosystems. The waste that is collected is either recycled (9%), incinerated, and/or combusted with energy recovery (5% in Canada and 12% in the US), or landfilled (80% or more).[2] The amount of plastic material in the environment exposed to full sunlight exceeds the quantity of submerged plastic (Royer et al., 2018).

In addition to the well-known contribution of landfills to the release of greenhouse gasses (particularly methane emissions), the creation of landfills to accommodate the increasing amount of plastic waste typically means destroying natural habitats for wildlife or the ecosystem functioning as a carbon sink (such as wetlands, parks, green corridors, etc.), directly impacting climate change mitigation interventions. Over 3,000 active landfills in the United States alone have contributed to 1,800,000 acres of natural ecosystem loss by 2021 (Vasarhelyi, 2021).

In addition to changes in land use, the production and release of toxic chemicals from the landfilled plastic have affected the health of the fertile soil and the entire landscape. Factors and conditions within landfills, such as light, heat, moisture, chemical oxidation, and biological activity, can cause changes in the physical and chemical structure of the polymer, resulting in the toxic composition becoming readily available to the environment (Royer et al., 2018). Particularly, the warming of the climate increases the rate of hydrocarbon production for terrestrial plastic waste compared to those in aquatic environments (Royer et al., 2018). If the leachate produced by plastic waste in landfill sites finds a pathway to the environment (e.g., the case with all open landfills), it contaminates nearby water sources, damages nearby ecosystems, and further exacerbates climate change. Duration of active post-closure operations of a landfill (known as the "aftercare" period) varies in countries around the world based on landfill conditions, but on average it is estimated

[2] Numbers indicate average statistics for US and Canada.

to be 140 years only after which the land is usable again (Heyer et al., 2005). A 2020 study of spatial patterns of mesoplastics and coarse MPs in floodplain soils has also indicated an overall widespread but spatial heterogeneous contamination in floodplain soils resulting from land use and fluvial processes (Weber and Opp, 2020). These adverse effects have been raised by indigenous communities globally for many years; notably, the Māori land guardians in New Zealand link plastic pollution to land and food sovereignty issues and believe that the trauma associated with the colonial nature of plastic pollution is similar to the climate change effects on indigenous oceanic territorial rights, economic injustice, and enduring imperial entitlement in the science and political sectors (Liboiron, 2020).

Conclusion

The crisis of plastic is closely tied to climate change. Over the last few decades efforts have been focused on reducing the use of fossil fuel resources as feedstock by replacing them fully or in part with renewable biomass resources, primarily to address the pollution problem in the ecosystem. With a global market of $17 billion in 2017, covering approximately 10–15% of the total plastic market and without a universally standard definition, unfortunately, what is widely known as "bioplastics" in North America and Europe present many of the same issues as traditional fossil fuel-based plastics (Bartolo et al., 2021). Defined as a plastic material that is either biobased, biodegradable, or features both properties, bioplastics are currently commercialized and are not only considered resource-intensive but are also for the most part rely on fossil fuels for production (European Bioplastics, 2016). The small proportion of bioplastics that are completely made from renewable materials typically results in perplexity during disposal and end-of-life management. In this category of materials, some are compostable which means they do not have any additive toxins that are left as a residue in the environment after biodegradation in a compost site. However, due to their different chemical composition, if mixed into the recycling streams, compostable plastics can contaminate the recycling system and reduce the value of recycled material, making recycling practices even less effective than it ever was (Cho, 2017). Moreover, due to limited infrastructure and industrial composting facilities that can produce the high temperature needed to break down compostable (or biodegradable) plastics, they are often disposed of in landfills causing them to be deprived of oxygen and may lead to releasing methane (Cho, 2017). That said, many researchers believe a new standardized generation of bioplastics has the potential to contribute to a circular economy and encourage further development of compostable bioplastics in targeting plastic packaging where recycling is challenged by food-contaminated layers (Kakadellis and Harris, 2020).

A bigger issue in the context of climate change is comparable GHG emissions during the lifecycle of alternative plastic products (Rosenboom et al., 2022). Figure 12 by Kakadellis and Harris (2020) illustrates an example of a flow diagram representing system boundaries during a lifecycle assessment. Considering a scenario wherein any fossil fuel-based plastic is identified as toxic material and completely banned from the global trades system, in 2010 the researchers at the University of Pittsburgh conducted a study that determined bioplastics produce much lower GHG

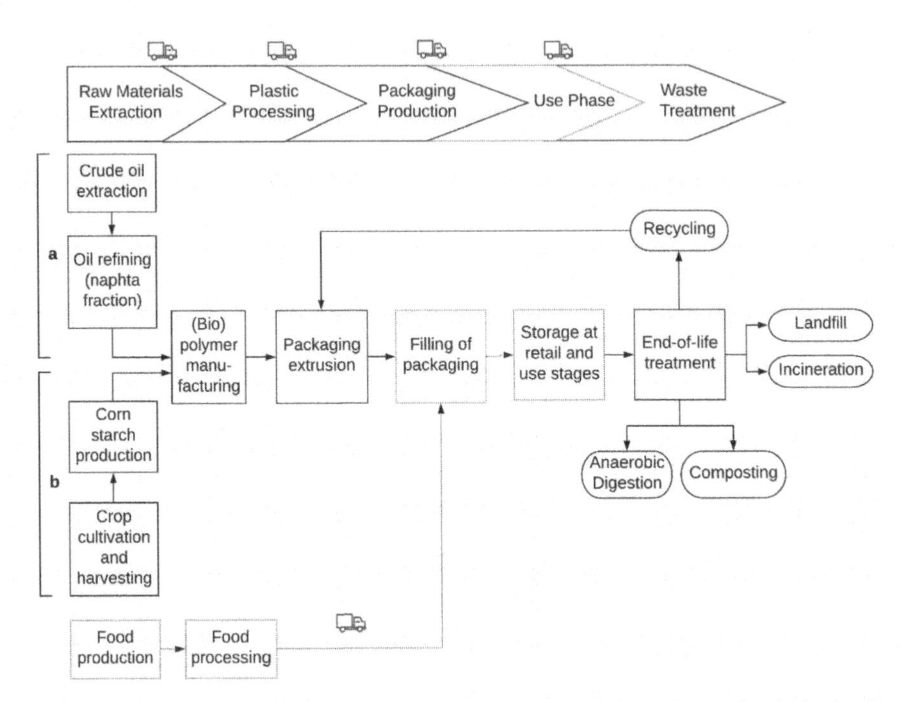

Fig. 12. LCA Flow Diagram Representing Packaging Process Using (a) Conventional Plastic; (b) Biodegradable and Bio-Based Bioplastic Material (Kakadellis and Harris, 2020).

emissions throughout their lifecycle due to the carbon dioxide which is offset during the growth phase of their raw materials, excluding the petroleum fuel used to run the farm machinery (Tabone et al., 2010). Further, switching from traditional plastic to corn-based polylactic acid (PLA) was shown to cut U.S. greenhouse gas emissions by 25% (Posen et al. 2017). This value can be further reduced by 50% to 75% if traditional plastics are still produced using renewable energy sources instead of fossil fuels (Posen et al., 2017). However, the production of bioplastics has shown a greater impact on ozone depletion and higher amounts of environmental pollutants due to the fertilizers and pesticides used during growing the crops as well as the chemical processing needed to turn organic material into plastic (Tabone et al., 2010). Hybrid plastics were also investigated and found to have the highest potential for toxic effects on ecosystems and the most carcinogens throughout their lifecycle due to the combination of the negative impacts of both agriculture and chemical processing (Cho, 2017).

In addition to comparable GHG emissions, bioplastics require extensive land use which competes with the land required for food production (Tabone et al., 2010). The Plastic Pollution Coalition (2017) projected that by 2019 more than 1.4 million hectares (3.4 million acres) of land would be needed to grow crops required to meet the growing global demand for bioplastics.

The problem with plastic is multifaceted and deeply interconnected with the changing climate. If addressed separately, each remains a competing problem within the policy space. The interconnection of the two issues needs to be integrated into the early stages of development to ensure joint solutions wherein solving one issue does not exacerbate the other. Many researchers around the world are developing novel production processes, alternative materials, or innovative business models to reduce the footprint of plastic substances and commercialize solutions without enormous GHG consequences. Many others propose policy directives that systematically prevent waste by disincentivizing the use of non-durable plastics and are independent of their feedstock or biodegradability claims (Plastic Pollution Coalition, 2017). More optimistically, global initiatives offer strong encouragement to existing and new climate actions by all societal actors. The inclusion of plastics in the agenda of the 26th United Nations Climate Change Conference of the Parties (COP26) in 2021 was a long overdue but positive step toward acknowledging the role of plastic in the global climate crisis. Nonetheless, reducing the climate cost of plastic needs far stronger global commitments. The United Nations Environment Assembly (UNEA) progress toward an internationally legally binding agreement to end plastic pollution is promising and fundamental in creating an evolving bottom-up polycentric system of governance to facilitate the introduction of a unique framework, policy interventions, and financial support for better management of conventional plastic in the economy, gradual reduction of plastic production toward full eradication, and ultimately restoration of our natural ecosystem.

References

Adyel, T.M. and Marcreadie, P.I. (2022, January 1). Plastics in blue carbon ecosystems: a call for global cooperation on climate change goals. Planetary Health 6(1): E2–E3.

Andrady, A.L. 2015. Persistence of Plastic Litter in the Oceans. pp. 57–72. *In*: Bergmann, M., Gutow, L. and Klages, M. (eds.). *Marine Anthropogenic Litter*. Springer, Cham. https://doi.org/10.1007/978-3-319-16510-3_3.

Aydın, C., Güven, O., Salihoğlu, B. and Kıdeyş, A.E. 2016. The Influence of Land Use on Coastal Litter: An Approach to Identify Abundance and Sources in the Coastal Area of Cilician Basin, Turkey. *Turkish Journal of Fisheries and Aquatic Sciences* 16: 29–39.

Bartolo, A.D., Infurna, G. and Dintcheva, N.T. 2021. A Review of Bioplastics and Their Adoption in the Circular Economy. *Polymers* 13(8): 1229. https://doi.org/10.3390/polym13081229.

BP. (2019, February 14). *BP Energy Outlook 2019 (Press Release)*. Retrieved from https://www.bp.com/en/global/corporate/news-and-insights/press-releases/bp-energy-outlook-2019.html

Cayan, D.R., Bromirski, P.D., Hayhoe, K., Tyree, M., Dettinger, M.D. and Flick, R.E. 2008. Climate change projections of sea level extremes along the California coast. *Climate Change* 87: 57–73. https://doi.org/10.1007/s10584-007-9376-7.

Cho, R. 2017. The Truth About Bioplastics. *State of the Planet*.

Dris, R., Gasperi, J., Rocher, V., Saad, M., Renault, N. and Tassin, B. 2015. Microplastic contamination in an urban area: a case study in Greater Paris. *Environmental Chemistry* 12(5): 592–599 https://doi.org/10.1071/EN14167.

Earn, A., Bucci, K. and Rochman, C.M. 2021. A systematic review of the literature on plastic pollution in the Laurentian Great Lakes and its effects on freshwater biota. *Journal of Great Lakes Research* 47: 120–133.

European Bioplastics. 2016. *Bioplastics Factsheet*. Retrieved from European Bioplastics: https://docs.european-bioplastics.org/2016/publications/fs/EUBP_fs_what_are_bioplastics.pdf.

Ford, H.V., Jones, N.H., Davies, A.J., Godley, B.J., Jambeck, J.R., Napper, I.E., Suckling, C.C., Williams, G.J., Woodall, L., Koldewey, H.J. 2021. *The Fundamental Links Between Climate Change and Marine Plastic Pollution*. Science of the Total Environment.

Galgani, F., Hanke, G. and Maes, T. 2015. Global distribution, composition and abundance of marine litter. Springer, Cham.

GESAMP. 2015. *Sources, Fate and Effects of Micro-Plastics in the Marine Environment: a Global Assessment.* Polestar Wheatons (UK) Ltd, Exeter, EX2 8RP.

Hamilton, L.A., Feit, S., Muffett, C., Kelso, M., Rubright , S.M., Bernhard, C., . . . Labbé-Bellas, R. 2019. *Plastic & Climate: The Hidden Costs of a Plastic Planet.* Center for International Environmental Law (CIEL).

Heyer, K.-U., Hupe, K. and Stegmann, R. 2005. Landfill Aftercare—Scope For Actions, Duration, Costs And Quantitative Criteria For The Completion. *Tenth International Waste Management and Landfill Symposium.* Italy: CISA, Environmental Sanitary Engineering Centre.

Hilmi, N., Chami, R., Sutherland, M.D., Hall-Spencer, J.M., Lebleu, L., Benitez, M.B. and Levin, L.A. 2021. The Role of Blue Carbon in Climate Change Mitigation and Carbon Stock Conservation. *Frontiers in Climate*, https://doi.org/10.3389/fclim.2021.710546.

Hou, L., Kumar, D., Yoo, C.G., Gitsov, I. and Majumder, E.L.-W. 2021. Conversion and removal strategies for microplastics in wastewater treatment plants and landfills. *Chemical Engineering Journal* 406, https://doi.org/10.1016/j.cej.2020.126715.

IEA. 2018. *The Future of Petrochemicals.* OECD/IEA.

IPCC. 2019. *Climate Change and Land.*

Jâms, I.B., Windsor, F.M., Poudevigne-Durance, T., Ormerod, S.J. and Durance, I. 2020. Estimating the size distribution of plastics ingested by animals. *Nature communications.*

Jambeck Research Group. (2015, February 12). *Plastic Waste Inputs from Land into the Ocean.* Retrieved from https://jambeck.engr.uga.edu/landplasticinput.

Jambeck, J., Geyer, R., Wilcox, C., Siegler, T., Perryman, M., Andrady, A., . . . Law, K. 2015. Plastic waste inputs from land into the ocean. *Science* 347: 768–771.

Kakadellis, S. and Harris, Z.M. 2020. Don't scrap the waste: The need for broader system boundaries in bioplastic food packaging life-cycle assessment e A critical review. *Journal of Cleaner Production 274*(https://doi.org/10.1016/j.jclepro.2020.122831), 122831.

Klaper, R. and Welch, L.C. 2011. *Emerging Contaminant Threats and the Great Lakes: Existing Science, Estimating Relative Risk and Determining Policies.* Alliance for the Great Lakes.

Kosior, E. and Crescenzi, I. 2020. Solutions to the plastic waste problem on land and in the oceans. pp. 415–446. *In*: T.M. Letcher. *Plastic Waste and Recycling.* London, United Kingdom: NEXTEK, Kensington Gore,.

Krantzberg, G. (2020, May/June). Plastic Pollution in the Aquatic Environment, Why it matters and what can we do about it. *Water Canada,* pp. 20–21.

Kuhn et al., S. 2015. Deleterious Effects of Litter on Marine Life. *In*: M. Bergmann and e. al, *Marine Anthropogenic Litter.* Springer.

Kvale, K., Prowe, A.E., Chien, C.-T., Landolfi, A. and Oschlies, A. 2021. Zooplankton grazing of microplastic can accelerate global loss of ocean oxygen. pp. doi: https://doi.org/10.1038/s41467-021-22554-w.

Mansouri, N.Y., Cham, R., Duarte, C.M., Lele, Y., Mathur, M. and Osman, M.A. (2020, November 22). Policy Brief: nature-based solutions to climate change: towards a blue carbon economy future. *Climate Change and Environment*, 1–23.

Marin, C. (2021, May). Global trade in plastic waste. *Le Monde diplomatique.*

Meeker, J.D., Sathyanarayana, S. and Swan, S.H. 2009. Phthalates and other additives in plastics: human exposure and associated health outcomes. *Biological Science*, https://doi.org/10.1098/rstb.2008.0268.

Napper, I.E., Baroth, A., Barrett, A.C., Bhola, S., Chowdhury, G.W., Davies, B.F., . . . Koldewey, H. 2021. The abundance and characteristics of microplastics in surface water in the transboundary Ganges River. *Environmental Pollution* 274: 116348. https://doi.org/10.1016/j.envpol.2020.116348.

NASA. (n.d.). *Global Phytoplankton Distribution.* Retrieved from My NASA Data: https://mynasadata.larc.nasa.gov/basic-page/global-phytoplankton-distribution.

National Research Council. 2010. *Advancing the Science of Climate Change.* Washington, DC: The National Academies Press. https://doi.org/10.17226/12782.

Ocean and Climate Platform. (n.d.). *The Role of the Ocean in Climate Dynamics.* ocean-climate.org.

Ocean Conservancy, McKinsey Center for Business and Environment. 2015. *Stemming the Tide: Land-based Strategies for a Plastic-free Ocean.*

Oehlmann, J., Schulte-Oehlmann, U., Kloas, W., Jagnytsch, O., Lutz, I., Kusk, K.O., Wollenberger, L., Santos E.M., Paull, G.C., Van Look, K.J.W. and Tyler, C.R. 2009. A critical analysis of the biological impacts of plasticizers on wildlife. *Biological Sciences*, https://doi.org/10.1098/rstb.2008.0242.

Plastic Pollution Coalition. 2017. *What is the Role of Bioplastics in a Circular Economy?* https://www.plasticpollutioncoalition.org.

Posen, D., Jaramillo, P., Landis, A.E. and Griffin, W.M. 2017. Greenhouse gas mitigation for U.S. plastics production: energy first, feedstocks later. *Environmental Research Letters.*

Rosenboom, J.-G., Langer, R. and Traverso, G. 2022. Bioplastics for a circular economy. *Nature Reviews Materials* 7: 117–137.

Royer, S.-J., Ferron, S., Wilson, S.T. and Karl, D.M. 2018. Production of methane and ethylene from plastic in the environment. *PLoS ONE* 13(8): e0200574. https://doi.org/10.1371/journal. pone.0200574.

Ryan, P.G. 2015. A Brief History of Marine Litter Research. pp. 1–27. *In*: M. Bergmann, M. Klages and L. Gutow, *Marine Anthropogenic Litter*. https://doi.org/10.1007/978-3-319-16510-3_1)). Springer International Publishing AG Switzerland.

Sebille, E.v., Aliani, S., Law, K.L., Maximenko, N., Alsina, J.M., Bagaev, A., Bergmann, M., Chapron, B., Chubarenko, I., Cózar, A., Delandmeter, P., Egger, M., Fox-Kemper, B., Garab, S.P., Goddijn-Murphy, L., Hardesty, B.D., Hoffman, M.J., Isobe, A., Jongedijk, C.E., Kaandorp, M.L.A, Khatmullina, L., Koelmans, A.A., Kukulka, T., Laufkötter, C., Lebreton, L., Lobelle, D., Maes, C., Martinez-Vicente, V., Maqueda, M.A.M., Poulain-Zarcos, M., Rodríguez, E., Ryan, P.G., Shanks, A.L., Shim, W.J., Suaria, G., Thiel, M., van den Bremer, T.S. and Wichmann, D. 2020. The physical oceanography of the transport of floating marine debris. *Environmental Research Letters 15*(023003).

Secretariat of the Convention on Biological Diversity. 2012. Impacts of Marine Debris on Biodiversity: Current Status and Potential Solutions. *CBD Technical Series No.* 67.

Stamper, M.A., Spicer, C.W., Neiffer, D.L., Mathews, K.S. and Fleming, G.J. 2009. Morbidity in a Juvenile Green Sea Turtle (Chelonia mydas) Due to Ocean-Borne Plastic. *Journal of Zoo and Wildlife Medicine* 40(1): 1960198.

Statistica. 2022. *Annual production of Plastics Worldwide from 1950 to 2020.* Retrieved from https://www.statista.com/statistics/282732/global-production-of-plastics-since-1950/.

Tabone, M.D., Cregg, J.J., Beckman, E.J. and Landis, A.E. 2010. Sustainability Metrics: Life Cycle Assessment and Green Design in Polymers. *Environmental Science & Technology.*

Thompson, R.C., Swan, S.H., Moore, C.J. and Saal, F.S. 2009. Our plastic age. *Biological Sciences* 364(1526), https://doi.org/10.1098/rstb.2009.0054.

Tolinski, M. 2012. *Plastics and Sustainability: Towards a Peaceful Coexistence between bio-based and Fossil Fuel-based Plastics.*

UNEP. 2020. *Addressing Single-Use Plastic Products Pollution Using a Life Cycle Approach.* United Nations.

UNEP. 2021. *Neglected: Environmental Justice Impacts of Marine Litter and Plastic Pollution.* United Nations Environment Programme.

van Reenen, C. 2020. *Our Fresh Water Is Seeing More Plastic Litter due to COVID-19.* IISD.

van Sebille, E., Wilcox, C., Lebreton, L., Maximenko, N., Hardesty, B.D., Franeker, J.A., Eriksen, M., Siegel, D. Galgani, F. and Law, K.L. 2015. A global inventory of small floating plastic debris. *Environmental Research Letters* 10: doi:10.1088/1748-9326/10/12/124006.

Vasarhelyi, K. 2021. *The Hidden Damage of Landfills.* Boulder: University of Colorado .

Wang, J., Liu, X., Li, Y., Powell, T., Wang, X., Wang, G. and Zhang, P. 2019. Microplastics as contaminants in the soil environment: A mini-review. *Science of The Total Environment* 691(15): 848–857.

Webb, J. 2012. *Climate Change and Society: The Chimera of Behaviour Change Technologies.* UK: sagepub DOI: 10.1177/0038038511419196.

Weber, C.J. and Opp, C. 2020. Spatial patterns of mesoplastics and coarse microplastics in floodplain soils as resulting from land use and fluvial processes. *Environmental Pollution, Volume* 267, https://doi.org/10.1016/j.envpol.2020.115390.

Welden, N.A. and Lusher, A.L. 2017. Impacts of changing ocean circulation on the distribution of marine microplastic litter. *Integrated Environmental Assessment and Management* 13(3): 483–487. https://doi.org/10.1002/ieam.1911.

World Health Organization. 2019. *Microplastics in Drinking Water.* License: CC BY-NC-SA 3.0.

Zapata, O. 2021. The relationship between climate conditions and consumption of bottled water: A potential link between climate change and plastic pollution. *Ecological Economics.*

Chapter 4

Plastics and Circular Economy

Greg Zilberbrant and *Eric Kassee**

INTRODUCTION TO CIRCULAR ECONOMY

Achieving sustainability in the way we live should be approached in many different ways in order to account for the complexities of modern life. What has once been seen as a miracle material for its versatility, processability, and low cost, is now generally seen as a villain that society should avoid and work to eliminate from our lives. Plastic products were originally created from a by-product of waste from the oil industry; their origin was directly tied to the smart re-purposing of waste products into valuable commodities. Decades later, we see how this miracle material has been drastically overused, and the result is pollution in our environment on an unprecedented scale that will take a great global effort to remedy. Recycling and purchasing choice helps in reducing plastic consumption, and single-use plastics are good options that should be pursued. But in order to create an economy where we can enjoy the benefits of plastic without the negative effects of pollution, we must change the way we treat the end-of-life disposal of plastic. This requires innovation on many fronts, such as designing products with end-of-life recycling and re-processing in mind and creating economic systems that allow for waste streams to be diverted into new raw materials. This chapter focuses on the circular economy with examples and case studies of how this can be achieved in practice today.

In recent years, the circular economy has reached a point of being common terminology among corporate leaders and policymakers alike. The Circular Economy Action Plan under the European Green Deal—the emergence of international standards initiated in the form of BSI 8001 and ISO 323 (British Standards Institution, 2017) (ISO, 2018)—provides the adoption of circular economy targets by multinational corporations such as IKEA that commit to being a circular business by 2030 (IKEA, 2020); even included the Tokyo Olympics in 2021 that used only recycled materials for medals and podiums (Euronews, 2021). There is no shortage of examples of efforts toward a circular economy at all levels of public and private organizations.

McMaster University, 1280 Main St. W., Hamilton ON L8S 4K1.
Email: zilberg@mcmaster.ca
* Corresponding author: ekassee@jbienv.com

There is also the convergence of key themes and strategies that could lead to an "effective circular economy". The two key terms of consideration are *effective* and *economy*. This chapter discusses the former further on but the latter, *economy*, is a critical term that can sometimes be overlooked when evaluating a specific material or product.

The *economy* in "circular economy" refers to the services that are provided to society. This is explicitly different from the concept of circularity which refers to the flow of material or group of materials. If the material is 100% circular (has 100% circularity), then it eliminates any potential waste. But this metric has nothing to say about the amount of time the product was in service, the amount of energy consumed to reprocess the material, the shipping requirements to a specialized facility to reprocess it (if not possible simply with equipment), or whether embodied resources, energy or carbon of the material is exponentially greater than the alternatives that could provide the same service. In short, circularity is one measurement of the effectiveness of a circular economy but not the only one nor the end state of the transition to a circular economy.

Figure 1 may be simplistic in its appearance, but it should not be mistaken as such. A linear system does not simply discard materials; a linear system discards the energy, craftsmanship, as well as cultural, social, and economic value. Figure 1 represents an entire systemic failure in preserving these, as well as other values, in the resources we obtain, the goods we create, and the services that are provided to society.

The challenge with simply looking at the material flows is that no activity is independent of another. Plastic bottles may be lighter transportation reducing fuel use, and plastic packaging may extend the shelf life of grocery products reducing food waste; plastic personal protective equipment may be the best way to manage the transition of a virus in a global pandemic.

The shift to a circular economy requires a focus on every aspect of our system. This includes obtaining raw materials, manufacturing goods, and servicing/extending the life of goods to managing the end-of-life of these goods in their initial and subsequent lives. The diagram below from the Ellen MacArthur Foundation, a global think tank focused on circular economy transition, highlights the numerous points of intervention that can occur rather than the disposal of material in a circular economy (Ellen MacArthur Foundation, 2017). The diagram is split into two sections, highlighting the distinct paths for "finite materials" versus "renewables" where the former can be returned to the natural environment, while the latter must remain in the system for continuous re-use.

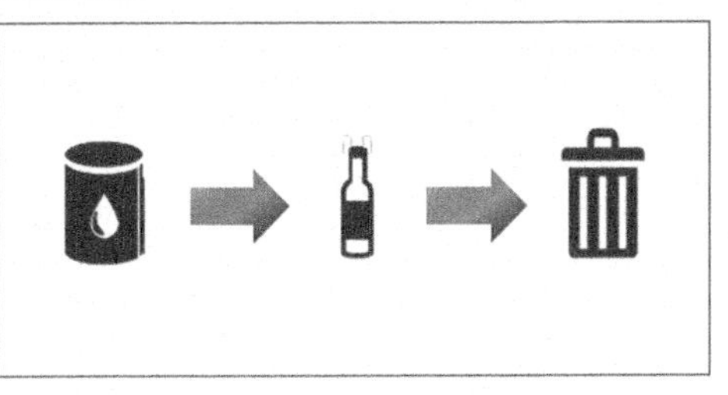

Fig. 1. Flow of Material in a Linear Economy—A Take-Make-Waste Process. Material is Extracted, Manufactured Into a Product for Consumption, and Eventually Discarded.

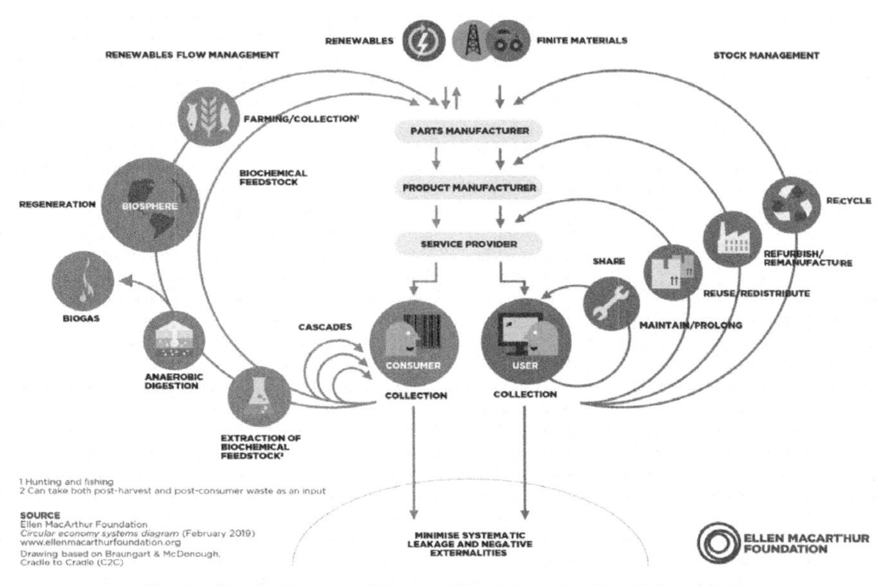

Fig. 2. Circular Economy Diagram (Ellen Macarthur Foundation, 2017).

An Effective Circular Economy

The effectiveness of the intervention is of critical importance when considering the transition to a circular economy. Efficiency can be achieved in a linear economy. The reduction of impact relative to the current level can be achieved without changing the way the system operates. The same quantities of material could be extracted, the same quantities of products or services provided, and with less environmental impact at the end of the process. Even extreme levels of efficiency, if possible, could reduce the amount of waste that goes to landfills and increase recycling rates or the emissions that enter the natural environment.

Perhaps one of the most fitting analogies of moving from efficiency to effectiveness was eloquently articulated by William McDonough (co-author of Cradle-to-Cradle) in *A Boat for Thoreau* (McDonough, 2000) which we will paraphrase below.

Consider embarking on a road trip from Toronto to Belleville, Ontario. This journey will require you to drive along Highway 401 for approximately two hours (200 kilometers), again highly dependent on traffic in Canada's most populous city. Although the road curves somewhat, at times, hug the shores of Lake Ontario, it is primarily northeast. In almost the exact opposite direction of London, Ontario, about the same distance from Toronto on the same highway, but is due southwest of Toronto.

The question very poignantly posed by McDonough (using American cities in his example) is: How long would it take you to reach Belleville when travelling from Toronto to London? If you were to slow down from 100 km/hr to 70 km/hr, a 30% reduction, or even 50 km/h, which is a 50% reduction, eventually you will still end up in London but never in Belleville. Now, London is a lovely town but if the intention is to go to Belleville, then you are simply headed in the wrong direction.

The point that McDonough is highlighting is that reducing the "speed" at which we create waste, release greenhouse gasses, or consume non-renewable resources, whether by 30% or 70%, will still result in our arrival at the same destination. It will simply take us longer to get there.

A change to the system—from a linear one to a circular one—is analogous to turning the car around. Focusing on where we want to end up as a society is more important than the speed at which we are travelling.

In order to do so, an *effective* circular economy must decouple material extraction, emissions and waste from social and economic progress. An *effective* circular economy operates within the natural boundaries of our planet for the benefit of society. And, if we are to benefit all of society on an ongoing basis, we cannot continue travelling in the same direction.

Plastics in the Circular Economy

Plastic is a blanket term used to refer to polymers of which there are many types with many different applications. It is useful to differentiate by resin type, as shown in Fig. 3, as they do not share the same processing, applications, physical properties, performance, and recycling characteristics with each other. Most single-use film packaging is made up of LDPE (low-density polyethylene) or HDPE (high-density polyethylene), and modern engineered plastics can have additive layers for enhanced performance characteristics, such as improved moisture resistance, adhesion, tear-resistance, and other physical improvements. These additives are incorporated during

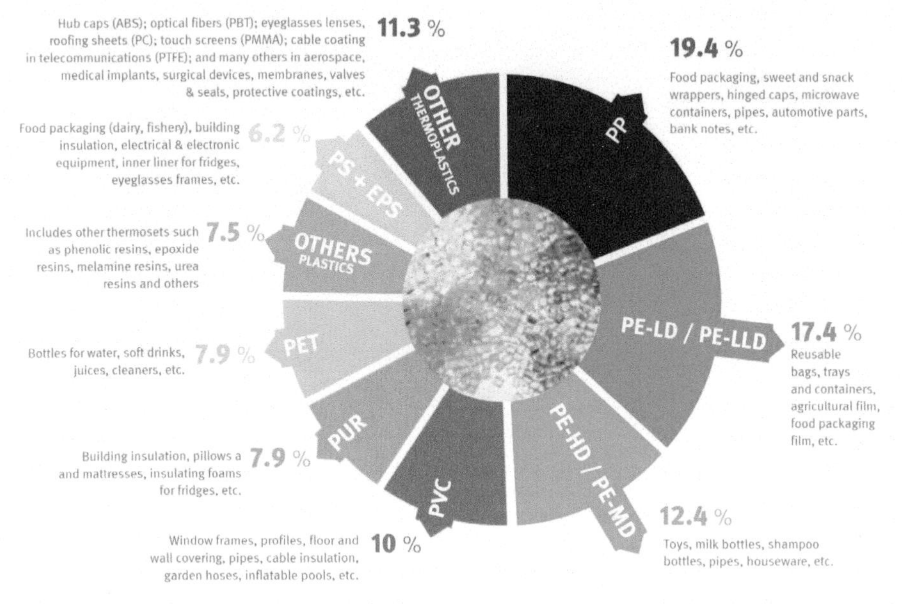

Fig. 3. Distribution of Plastics Demand by Type and Application (PlasticsEurope, 2020).

the extrusion stage of production and are nearly impossible to physically separate from base polymers after this step, which leads to difficulty in recycling. Other resins are often chosen for ease of production and the properties desired in final products. PET (polyethylene terephthalate) expands easily and durably when heated, so it is often used in blow moulding for high output production of plastic beverage containers. Polystyrene (Styrofoam) expands easily and becomes rigid, making it useful in packaging fragile items, while also providing temperature insulation. Poly-vinyl chloride (PVC) is strong and resistant to chemicals, so it is often used in plumbing and in applications that are intended to last for a long time. There are many applications for all types of resins, and even more when combined to optimize specific properties.

Plastic, in many of its forms, could be argued as being incompatible with the circular economy. That the material itself cannot exist in a functional circular economy. However, this assessment is usually based on behaviour and the category of plastics referred to as single-use or nondurable plastics. Materials, such as plastic shopping bags or drinking straws, have garnered most of the attention with plastic bans focused on these materials in Canada alongside stir sticks, six-pack rings, cutlery, and hard-to-recycle take-out containers (European Commission, 2021; Government of Canada, 2020).

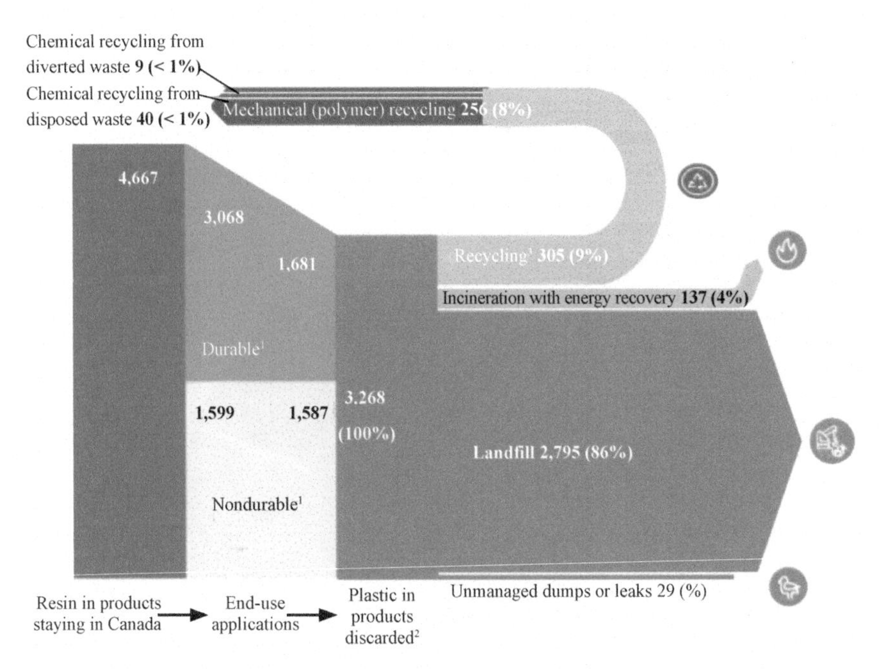

[1] Durable applications with an average lifetime > 1 year will end up as waste only in later years; given market growth and increase share of plastics in durable applications (e.g., construction, cars) plastics waste generated today is less than what is being put in the market that same year. On the contrary nondurable applications go almost straight to waste.

[2] 1,587 thousand metric tons of mixed plastic waste from nondurable applications 1,681 thousand metric tons of mixed plastic waste from production in previous years.

[3] Output recycling rate, after taking into account process losses.

Fig. 4. Canadian Resin Flow in Thousands of tonnes Annually (2016).

Nondurable plastics represent the largest portion of plastic waste in both Canada (as shown in Fig. 4) and the EU, at 47% and 50%, respectively. The same recycling systems that struggle to recycle 9% of plastic waste in Canada are not optimized to efficiently repurpose the types of plastic used in higher-end applications, such as for electronics, furniture, and other textiles. These systems have been optimized for the largest plastic waste-generating sector, which is packaging (ECCC, 2019). Even when the collection is optimized, the nature of the material (either in very low quantity or in highly heterogenous form or both) does not translate into material that will be reintroduced back into the system for new materials or products (ECCC, 2019).

The construction sector represents the highest use of durable plastics and the second highest in overall production with 26% of plastic utilization in Canada, and 20% in Europe (in terms of utilization, construction is a close second to packaging which represents 33% of use in Canada). However, this sector represents only 5% of plastic waste produced in Canada (ECCC, 2019). In the construction sector, plastics can often be the best choice for a long-lasting material, offering durability and longevity improvements over alternatives. It is used for moisture and vapour barriers, insulation, and piping where its performance specifications and lower cost are advantageous. In this case, plastic products have a useful life of decades as opposed to minutes with many single-use items. Plastic has only been widely used in construction since the 1980s in Canada, which also plays a part in the low waste numbers for this sector as most of the plastic is still actively serving its purpose in constructed buildings. However, the plastic waste numbers will increase as buildings need to be renovated or refurbished, so durable plastics are not immune to ending up in the same situation as their nondurable counterparts without proper intervention.

Despite recycling efforts, an excessive amount of plastic is ending up in a landfill and the environment. This reflects a fundamental flaw in the approach when looking at the end of the process to find solutions. Of the estimated 8.3 billion tonnes ever produced between 1950 and 2015, nearly 60% of that is estimated to exist in either the natural environment as pollution or landfills; 12% of it is estimated to have been incinerated, and only 9% has been recycled globally with the remaining 19% of plastic ever produced still in use today (Watkins, 2019).

So if plastic production was to cease today, plastic recovery efforts in the form of policy intervention and technological innovation must continue. Globally, approximately 5 billion tonnes of plastic material is either in the natural environment or in a landfill (UNEP, 2018) with an additional 1.6 billion tonnes currently in service that may suffer the same fate as the other 5 billion tonnes without proper action. However, this is not the reality as the plastic demand remains at an all-time high with over 300 million tonnes produced annually.

This speaks to a problem that requires a multi-pronged solution. Namely, 1) valued products and services need to be redesigned to eliminate plastic pollution and waste; 2) plastics destined for disposal or already in the natural environment need to be recovered; and 3) policies must be established to incentivize actions that eliminate or recovers waste plastic and clearly disincentivizes the opposite.

Plastics, like all materials, provide a service to society and the ability of plastics to be compatible with a circular economy requires the application of the same building blocks as any other materials. The next sections describe the key building blocks for a circular economy with a focus on the effective application of these building blocks to achieve the aforementioned multi-pronged solution, demonstrated with case studies in the plastics industry.

Building Blocks of a Circular Economy

The building blocks of a circular economy can be simplified into a few key approaches and prioritized accordingly. Figure 5 below shows the key approaches to transition to a circular economy with the three central approaches avoiding the potentially linear outcomes, while the outer rings aim to minimize existing impact with less and less desired results. The outermost ring represents an endpoint that is the same as a linear economy disposal scheme.

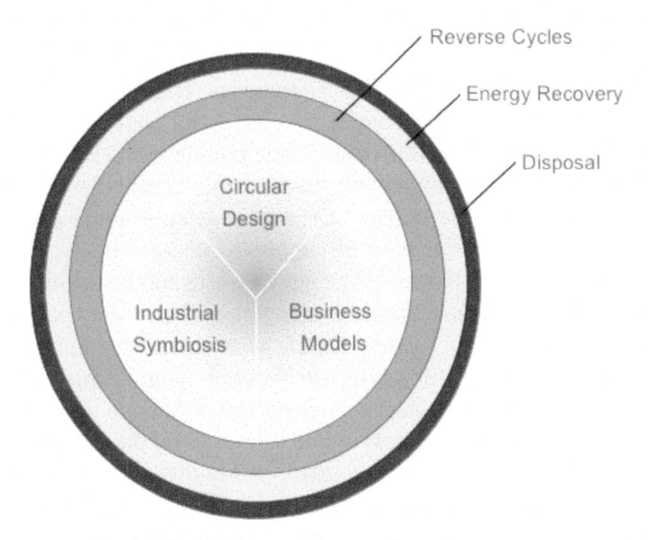

Fig. 5. Circular Economy Interventions.

The following sections further examine these approaches and how each can contribute to a circular economy.

Circular Design

Circular Design is an approach that integrates circularity into the design of a product or a material from the onset. The outcome of this approach is that no waste is produced. All non-renewable materials can be disassembled and repurposed, while any renewable or biological materials can be returned to the natural environment without a negative impact. This results in materials being reintroduced in the manufacturing of a similar new product, reprocessed into their original primary ingredient, or bio-based materials returning to the natural environment.

We have many examples of circular designs with precious metals. Gold, silver, and other precious metals can be re-smelted for the production of new products. This is especially true of jewellery where the metals are not mixed with other materials,

Circular Design Case Study: Ecovative

Ecovative directly reduces the amount of plastic and Styrofoam waste being produced and discarded by providing packaging without waste. Their mycelium technology is offering a new method for securely packaging fragile and oddly shaped products for shipment to consumers.

Using just two ingredients, hemp hurds and mycelium, they are able to custom grow a durable package structure to surround and protect a product during shipping which traditionally required styrofoam, plastic, air, or cardboard to help reduce impacts and damage. Their process involves nine steps to creating a durable package for shipping: computer automated design (CAD) to determine required dimensions, CNC milling, and tooling of MDF product analog, thermoforming growth trays to the dimensions of the MDF part, filling the thermoformed mold with hemp hurds, mycelium, and other composite and reinforcing mediums when desired. Then the parts are grown internally in pods, popped out of the trays for additional external growing, and placed into a chamber for drying. The result is a thermally insulating and water-resistant package that offers protection for fragile items during shipping which can then be safely disposed of alongside other compost. This growth process takes seven days, while the end packaging will biodegrade when added to soil in 45 days.

The need for a truly biodegradable option has been addressed with this design, rather than addressing the need for a recycling plan. Creating a product that can easily and safely be returned to the natural environment by end consumers shifts a lot of the burden away from municipal recycling programs and simplifies the extended producer responsibility for the shipped product as the alternatives (mainly styrofoam and blister packaging).

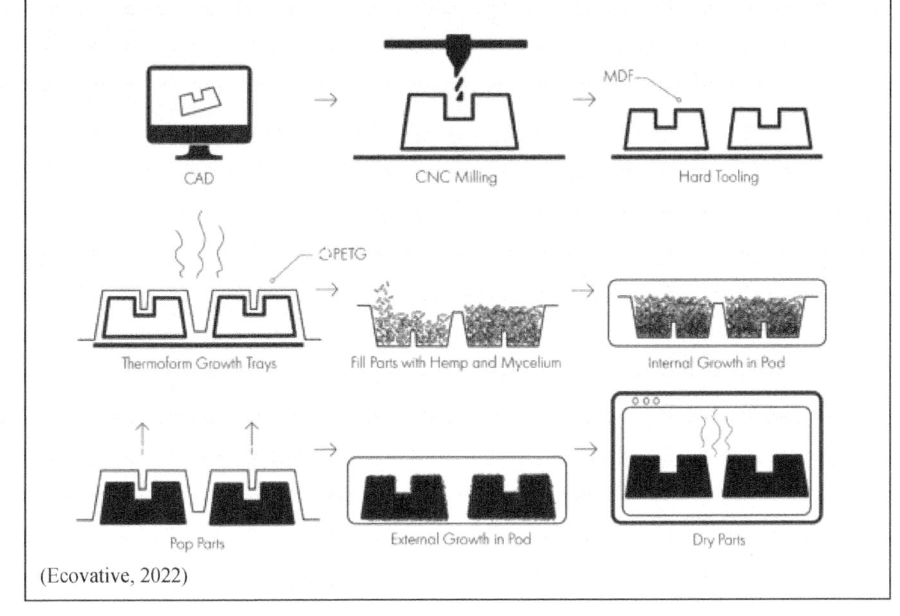

(Ecovative, 2022)

so the reprocessing does not produce waste by-products any more than would be produced by using mined ore. In fact, the purity of the secondary material results in less waste and resource usage than processing ore (Fothergill, 2004). The same concept applies to other metals such as aluminum and steel.

The circular design is intentional from the onset. Circular design can be more broadly adopted at a system level if organizations actively seek to reuse their materials and more importantly design their products to facilitate that reuse. The fundamental aspect of a circular design is that the producer has created something that has value beyond its initial use, whether it is to the organization to reintroduce into their process or the market or to the natural environment as nutrients.

As the name implies, waste is "designed" out of the system. This requires reflection on the purpose of the service being provided. Southwest Airlines former President Colleen Barrett famously stated, "We're in the customer service business, we just happen to fly airplanes" (Southwest, The Magazine, 2015). Just as intelligent oil and gas companies must realize that they are in the business of energy and focus on renewable energy to ensure their organizations are compatible with a sustainable future. Plastic and chemical companies must realize that they are in the business of packaging, food preservation, and other critical services to society that plastics currently provide in order to be compatible with a circular economy.

With that mentality, the purpose of the service informs the design of the product. Mushroom-based packaging like Ecovative is a great example of bio-based circular design. The material is made from food waste and mycelium (the fungal network that mushrooms sprout from) to create a customer-shaped material to protect fragile products in shipping; a role traditionally performed by plastic products, Styrofoam.

Industrial Symbiosis

Industrial symbiosis refers to the beneficial exchange of materials between industrial operations. Most commonly, it is the use of one industrial process' by-products as the feedstock for another industrial process. It can be seen as the intelligent design of the processes of multiple compatible industries to minimize ecological harm and maximize economic benefits through a material exchange.

A great example of industrial symbiosis is the partnerships between the steel and cement industries. Blast furnace slag, a by-product of iron production, is utilized by cement manufacturing to supplement traditional Portland cement directly or as raw material feed. The slag can be utilized as a replacement for Portland cement which is the traditional material manufactured by cement plants. By utilizing this by-product from the steel industry, the cement industry reduces their dependence not only on the raw material but can also bypass the thermal (kiln) process needed to produce clinker—the intermediary product used to make cement—thereby reducing the fuel consumption and emissions associated with manufacturing this cement substitute.

Broader examples of industrial symbiosis can exist as eco-industrial parks, a collection of industries that exchange resources. The most notable example is the eco-industrial park in Kalundburg, Denmark, where over a dozen facilities exchange material, water, and energy resources in a symbiotic relationship that supports the needs of each facility. Although Kalundburg is rightfully cited as a model for industrial symbiosis, effective industrial symbiosis can be achieved on a smaller scale and incorporate the strategy of Reverse Cycles that will be presented later in this chapter.

One company exemplifying industrial symbiosis in the plastics industry is C.R. Plastic Products. Located in Stratford, Ontario, C.R. Plastic Products utilizes plastic water bottle caps, milk jugs, and juice containers in the manufacturing of outdoor furniture. The use of HDPE recovered from their subsidiary recycling company, rather than virgin HDPE feedstock, reduces the need for virgin raw material for the furniture manufacturing business. Although a simple example, what separates this notion of industrial symbiosis from simply being referred to as recycling is the purpose for which it is executed. The purposeful use of by-products as raw materials requires a relationship (however, formal or informal) between the industry generating the by-product and the industry utilizing it as raw material.

Industrial Symbiosis Case Study: CR Plastics

CR Plastic Products is helping make the lifecycle of plastics more circular in their production of outdoor furniture products in Stratford, Ontario. They utilize high-density polyethylene (HDPE) recovered from recycling plants and turn them into long-lasting durable outdoor furniture products, such as deck chairs and tables. This allows for a seamless flow of recycled material to be processed (shredded, remelted, extruded, and reformed) into multiple products based on the type of material recovered.

Diverting waste from landfills becomes much easier and more desirable when there are viable and profitable alternate uses for this discarded material. CR Plastic Products has developed a sustainable business model that operates on two important circular economy principles i.e., finding a profitable use for material that would otherwise be discarded and creating a new product that is durable and will serve its intended purpose for a long time; at which time its components can be recycled and reprocessed again into another new and valuable product.

(C.R. Plastic Products, 2021)

Industrial symbiosis can lead to overall reductions in virgin raw material and energy use through these beneficial exchanges. It is the manifestation of "waste = food", where the waste of one process becomes the food (read: raw material) for another.

Circular Business Models

Filling a need in society is often heralded as the rationale to start a viable business. As the problem of plastics pollution is well documented, the emergence of new business models provides an opportunity to meet a need in society that does not create or contribute to the problem. Sourcing raw materials from the economy, rather than from raw ecological reserves, is one method as is repairing a product destined for a landfill. Even repurposing a discarded or seemingly end-of-life product from the energy-intensive transport and recycling processes is a value-adding niche that can be undertaken by any aspiring and entrepreneurial organization.

With the well-deserved negative publicity that single-use plastic products get, as well as the low historical recycling rates around the world, a new circular business model has shown a viable path to reducing the amount of plastic waste i.e., refillery services.

Refillery services are creating a circular business model by applying this concept to an increasing variety of common household products that have often been sold in single-use plastic containers.

By offering customers a way to simply refill their empty glass or durable plastic bottles with products, such as shampoo, dish soap, detergent, and even toothpaste, they eliminate the waste generated by end-of-life disposal of these products' traditional packaging. The refillery model even reduces the burden of recycling plastic or glass, as you can clean and reuse these containers, as long as they remain structurally sound. Even refilling a shampoo bottle once saves another bottle from needing to be recycled or ending up in a landfill. Refillery services exist in many forms, such as a booth at a grocery store, entire locations dedicated to supplying products to consumers using customer-supplied containers, or mobile services that operated with regular pop-up locations.

This business model achieves circularity by eliminating the need for individually packaged items. It delivers a product directly to the consumer which bypasses the need for packaging designed to last for long periods in warehouses or survive shipping around the world. By redesigning the entire service model, even at a small

Circular Business Models Case Study: Refillable Household Cleaning Products

Refillable Household Products

In many instances of single-use plastic, the sole purpose of the plastic is packaging to preserve a product until it is delivered to the end consumer. Finding alternative ways to obtain these products and preserve them from their production through their delivery to the consumer and throughout their active-use lifetime without any end-of-life waste is a circular economy priority. There are many products and companies that offer alternative ways to access common household products while avoiding the waste of unnecessary packaging. In Ottawa, Ontario, grocery chain *Terra20* has created *Ecobar,* Canada's largest refill station. This system allows customers to buy cleaning and personal care products from well-known sustainable brands, return the bottles when finished for cleaning and reuse, and receive a discount on subsequent refills. This process directly reuses the plastic bottles, instead of disposing of them in a recycling center.

(Terra 20, 2021)

The Glass Jar Refillery in Hamilton, Ontario, offers similar refill options to in-store customers with the benefit of being able to bring in and fill their own (pre-cleaned) containers.

(The Glass Jar Refillery, 2022)

Another choice available to consumers is the mobile refillery option. Halton Region in Ontario-based refillery market has no brick-and-mortar location. Instead, they use their website to allow customers to view and order products for next-day delivery and refilling of personal containers (Refillery Market, 2021).

In Burlington, Ontario-based Park Market & Refillery has a combination of a storefront and a mobile refill van that will also drive to customers' locations to fulfill orders.

(Park Market and Refillery, 2022)

scale, significant waste reductions can be realized when focusing on the consumer need and the cleaning products rather than the packaging that contains them. As such, circular business models have a critical role to play in delivering value to society without the environmental burden.

Reverse Cycles

"Reverse cycles" is perhaps the most exciting opportunity for a circular economy and the area where policy can accelerate the transition. This approach is one where organizations actively seek out materials that have been discarded to introduce them back into the manufacturing process in place of virgin materials. Whether in an industrial stockpile, in waste bins, in communities, or in the oceans, reverse cycles present an opportunity to create economic and social value by literally bending the last stage in the linear economy arrow (disposal) in the other direction. Organizations, like Plastic Bank and Interface, are actively seeking to create a solution for waste materials in the natural environment.

Plastic Bank, a Canadian-based organization, incentivizes the recovery of discarded plastics in developing nations. By monetizing the recovery of plastics for all community members, Plastic Bank is able to engage people to collect and return waste plastics from their communities to one of the company's recycling centers. In turn, the company sells the cleaned, sorted material to large plastics processors and more critically returns material into the system that was otherwise lost.

Another organization, Interface, is a recognized sustainability leader for decades and has been taking plastic recycling one step further. The company relies heavily on Nylon-6 for the manufacturing of its carpet tiles; a material that can be recycled from old carpets or purchased as raw feedstock. Nylon-6 is also the same material that is used in manufacturing fishing nets. Unfortunately, this material is present in any

Reverse Cycles Case Study: Plastics Bank

Plastic Bank is an organization that seeks to reduce plastic waste by monetizing the value of discarded plastic products and using them as feedstock to make recycled plastic products. This incentivized collection reduces the amount of plastic ending up in waterways and landfills and creates a circular economy of plastic collectors and producers.

Discarded plastic is collected by individuals and brought to centralized collection and distribution centers. The material is weighed, sorted, and the collectors are paid with tokens based on the value of the material collected. These tokens act as a type of currency and can be used to purchase groceries, fuel, school tuition, health insurance, and more. This conversion of a polluting waste product into a valuable part of a supply chain has dramatic effects on participating communities, giving them a sustained source of income where there was none before. Once the plastic is collected and paid for, it is processed for reuse by washing, flaking, and separating. Integrating this process with a blockchain network allows the entire process to be secure, provides real-time data visualization, and makes the process transparent, traceable, and scalable.

The collected and processed plastic is then turned into *Social Plastic* which is a raw plastic supply that can be used in place of virgin resins created from fossil fuels. Creating products out of social plastic is a significant contributor to a closed-loop supply chain. This technologically-enhanced new recycling ecosystem reduces ocean plastic waste and the reliance on virgin plastic while supporting communities that can benefit from this new resource.

(Plasticbank, 2021)

natural environment where there is fishing activity due to loss during fishing, tears/damage, and discarded nets (Net-works, 2021). By actively engaging small fishing communities, Interface has created a system where damaged or discarded fishing nets can be sold to partner organizations for use in the production of their carpet tiles.

As with Plastic Bank, Interface is actively seeking to return material that has been discarded, while creating economic benefits for the communities. This approach integrates secondary material as feedstock into their supply chains and returns valuable resources to the economy. Reverse cycles aim to correct the failings

Reverse Cycles Case Study: Interface Net-works

Interface has been a model for sustainable manufacturing since its Mission Zero commitment in 1994, and its passion and ability to find innovative solutions to production sustainability challenges have continued to evolve since. *Net-Works,* an initiative supported by *Zoological Society of London*, by *Interface* is an inclusive business model that combines plastic waste removal and prevention with marine life preservation and economic opportunities.

Net-Works was originally trialed in 2012 in the Philippines Bantayan Islands and has since grown with hubs being established in Danajon Bank and Cameroon. This program is an all-encompassing solution to a few different but related problems in coastal fishing communities. Nylon-6 fishing nets often break after significant use and have traditionally been discarded in the water which has devastating effects on marine animal and plant life. Conservation efforts in the past can only do so much to help the problem, but fishing nets are disposed of at such a rate in the world's oceans that a year's worth of nets stretched end to end could wrap around the entire planet 1.39 times. *Net-Works'* solution to this problem is to provide real incentives to local fishing communities to recover these nets and dispose of them properly in the first place. Between 2012 to 2015, 66,860 kg of fishing nets were collected. These nets are then dried and bailed for shipping where they are then sold to nylon yarn supplier *Aquafil*. Having this market to sell discarded nets to is crucial to the success of this project as financial empowerment is a driving force of this successful buy-in by community participants. It is estimated that over 2 billion people in the world do not have a bank account which isolates them from many opportunities for school, housing, and retirement. Selling these nets is a way to bring in a consistent income which also brings with it many opportunities for socioeconomic improvement.

Once *Aquafil* receives these nets, they recycle them into nylon yarn which is then bought by *Interface* to be used in their carpet tiles. This industrial ecology that was born out of a need to reduce plastic waste in the oceans has turned into a sustainable recycled carpet supply chain and as a result, it has many financial and socioeconomic benefits to impoverished coastal communities.

(Net-works, 2021)

of linear action by the proceeding users of material and the actions of predecessors that chose a linear approach which, as mentioned earlier in the chapter, discards the energy and craftsmanship, as well as the cultural, social, and economic value.

There are many more solutions out there that have been proven to work and are not only cost-effective but can stimulate economic progress. Plastic pollution in our environment is not inevitable. Diverting existing sources of pollution by means of a circular economy and recovering the current plastic littering our planet are two equally important aspirations. The shift from seeing discarded plastic as garbage and waste to treating it as a useful resource is how we shift linear plastic supply chains into becoming circular. By incentivizing recovery and proper disposal/recycling instead of incentivizing continual waste production, we not only fund a solution to pollution but also create the conditions for a circular economy to thrive.

Energy Recovery & Disposal

Energy Recovery remains the last ring of the circular economy—and arguable is more an element of a linear system than a circular one. It remains a viable tool for waste management to prevent dumping or landfilling of materials that will create impact to the nature environment. The material value is ultimately lost but the energy value is recovered.

The problem of course, is the emissions created in the combustion of waste materials whether this takes places in an energy-from-waste facility generating electricity or as an alternative fuel in energy-intensive industries such as cement manufacturing. Energy generation and reducing the volume of waste in landfills would be an obvious win-win solution, if not for the emissions that result from the combustion process. In some energy-intensive industries, the use of plastics as fuel may be substituting coal or other fossil fuels that can lead to a reduction in certain emissions from combustion as well as avoiding extraction of fossil fuels to support the high energy needs of these processes. However, from a broader perspective—this remains a tool of a poorly designed (linear) system but a necessary one in the transition, if the only alternative is disposal.

Disposal can take the form of landfilling or thermal destruction (without energy recovery). Formal waste management systems remain an important part of modern infrastructure, whether as public or private services, to avoid dumping or uncontrolled burning. Whether is in the form of energy recovery or disposal—the destruction of embodies resources is inconsistent with the goals of an effective circular economy. This necessity is a stark reminder of the failed design of the linear economy, however, the opportunity to redirect materials from landfill and energy recovery to higher value uses is a role that traditional waste management operators can play to support the transition to a circular economy.

Policies Supporting the Circular Economy Transition for Plastics

Reaching the goal of an effective circular economy, as logical as it is from a social, environmental, and economic standpoint, still requires the proper policy environment and instruments to facilitate the transition. Two common policy approaches are Extended Producer Responsibility (EPR) and Deposit Return Systems (DRS).

EPR is a policy instrument that refers to regulations that make the producers of a product responsible for its materials' entire lifecycle, including recovery and recycling of end-of-life material components and even the packaging used to deliver the product to the consumer. Most OECD countries have at least some type of EPR regulation even if it extends only to specific sectors, such as packaging, batteries, or electronics (OECD, 2014). Throughout Canada, there currently exists over 120 individual EPR programs across the provinces and territories (EPR Canada, 2017).

Around the world, EPR adoption has been growing for years. The collection, sorting and recycling of a package typically cost more than the value of the material recovered, so EPR is the only way to fund these recycling efforts sufficiently (Ellen Macarthur Foundation, 2022). EPR is slowly but surely spreading around the globe with more than 25 countries in Europe having at least some type of EPR legislation, mostly related to household plastics (EXPRA, 2016). Figure 6 shows the growth in packaging EPR legislation around the world since 2000.

While having legislation levels the playing field for companies operating across jurisdictions, individual businesses and organizations have their part to play, and they are also realizing that EPR policies are becoming more of a necessity. The Ellen Macarthur Foundation has reported that over 150 leading businesses and organizations have publicly recognized the need for EPR, and over 500 companies have signed commitments to reach targets for achieving a circular economy for plastics (Ellen Macarthur Foundation, 2021).

PACKAGING EPR IN 2000 ■ In effect

PACKAGING EPR IN 2018 ■ In effect (mandatory & voluntary) ■ In implementation ■ Framework EPR legislation

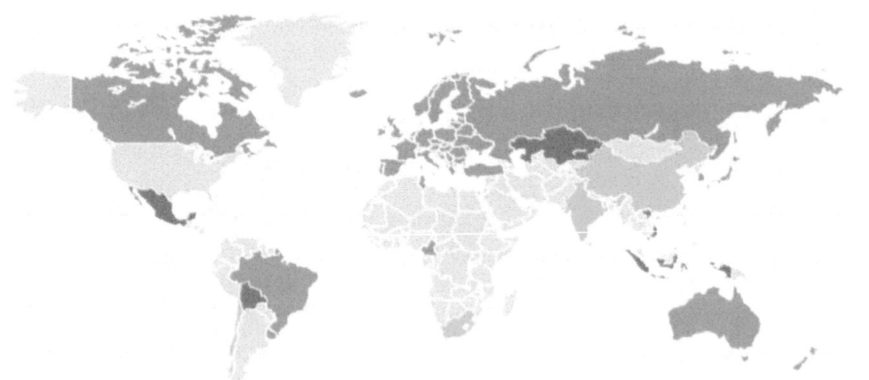

Fig. 6. International Growth of Extended Producer Responsibility Policies From 2000 to 2018 (WWF, 2020).

Through EPR the economic burden of end-of-life material management systems is placed on the producer, creating the financial incentives to improve the design. The consumer can also be incentivized to help manage the destination of recyclable plastics. A proven, effective way to encourage the proper management of plastics and recyclable items is deposit return systems (DRS).

DRS refers to an incentivized recycling collection system targeted directly at consumers. Retailers charge a small fee (usually a few cents) on the upfront sale of eligible products, such as plastic, glass, and aluminum beverage cans and bottles. The fee is then refunded to the consumer when the empty container is returned to either the same location or designated retailer or a centralized recycling and sorting center. This reduces operating costs and improves the uptake and efficiency of the recycling and collection system.

Deposit return systems around the world have proven to work at reducing waste and encouraging consumers to make the choice and effort to recycle with often dramatic results. Norway introduced a deposit return system in 2018. Consumers pay a small deposit fee embedded in the cost of a product up front and later return the bottle or other empty package and collect the fee for their deposit at a centralized recycling centre or the more convenient option of a "reverse vending machine" which are available outside many retailers, such as grocery stores. These machines accept a large variety of bottle shapes and materials and will calculate the value of the deposit and print a coupon with a cash value that can be redeemed in the store. This system has resulted in Norway having a 97% recycling rate for plastic bottles, putting it well ahead of the European Union's 2029 deadline to achieve a 90% recycling rate (Deshayes, 2020). Another successful example is Lithuania which improved a 34% recycling rate to over 92% within 2 years of implementing DRS policies (INNOWO, 2020; USAD, 2022).

Incentivizing Deposit Return Systems in combination with Extended Producer Responsibility and minimum recycled content policies can further support the building blocks of a circular economy. The building blocks of a circular economy for plastics are not independent of one another, and they in fact rely on each other and are strengthened by the symbiotic relationship of the system being created. This is supported by policies that create the environments to accelerate the adoption of these building blocks. Reducing the dependence on virgin plastics is critical but the current market demand for the services provided by plastics must then be met by some other means. The transition to an effective circular economy requires assessing the needs of our society, designing solutions that meet those needs, and ensuring the resulting solutions do not waste material, energy, cultural, social, and economic value. The future role of plastics in our society will be determined not by their chemical composition but by their adherence to the building blocks of an effective circular economy.

The processes and methods discussed in this chapter serve as not just examples of effective action on plastic pollution but also as models of innovative thinking and creative solutions to this difficult problem. Since global supply chains and many products that are part of daily life rely on plastics right now, eliminating plastic is much more complex than just stopping the use and production of single-use plastics. Some of the solutions presented are clever ways that allow traditional packaging and shipping to continue but without end-of-life waste, such as the mycellar packing example. Other innovations, such as refillable product containers or Net-works, reduce end-of-life waste by reprocessing the waste before it enters the environment. These multi-faceted approaches, targeting all areas of the plastic lifecycle, will make the work of end-of-life repurposing of a material a much more manageable task. These combined approaches and examples of creative thinking are how we as a society can have a measurable shift toward a circular economy for plastics.

References

Bloxsome, N. (2020, October 20). *International Aluminium Institute publishes global recycling data.* Retrieved from aluminiumtoday.com: https://aluminiumtoday.com/news/international-aluminium-institute-publishes-global-recycling-data.

British Standards Institution. (2017, May). *The rise of the Circular Economy.* Retrieved from bsigroup.com: https://www.bsigroup.com/en-GB/standards/benefits-of-using-standards/becoming-more-sustainable-with-standards/BS8001-Circular-Economy/.

C.R. Plastic Products. (2021, July). *Our Commitment to the Environment.* Retrieved from crpproducts.com: https://crpproducts.com/about.html.

CCSP. 2020. *2020 Annual Report on the State of Sustainable Public Procurement in Canada.* Vancouver: The Canadian Collaboration for Sustainable Procurement.

Deshayes, P.-H. (2020, February 13). *In Norway, bottles made of plastic are still fantastic.* Retrieved from phys.org: https://phys.org/news/2020-02-norway-bottles-plastic-fantastic.html

E. Kosior, I.C. 2020. Solutions to the plastic waste problem on land and in the oceans. *Plastic Waste and Recycling: Environmental Impact, Societal Issues, Prevention, and Solutions*, 415–446. doi:doi.org/10.1016/B978-0-12-817880-5.00016-5.

ECCC. 2019. *Economic Study of the Canadian Plastic Industry, Markets and Waste.* Deloitte, Cheminfo Services. ECCC. Retrieved from https://publications.gc.ca/collections/collection_2019/eccc/En4-366-1-2019-eng.pdf

Ecovative. (2022, January). *We grow better materials.* Retrieved from ecovative.com: ecovative.com

Ellen Macarthur Foundation. 2017. *Infographic: Circular Economy System Diagram.* Retrieved from ellenmacarthurfoundation.org: https://www.ellenmacarthurfoundation.org/circular-economy/concept/infographic

Ellen Macarthur Foundation. 2021. *Global Committment: Signatory Reports.* Packaging producers and users. Retrieved January 11, 2022, from https://ellenmacarthurfoundation.org/global-commitment/signatory-reports

Ellen Macarthur Foundation. (2022, January 12). *Extended Producer.* Retrieved from https://plastics.ellenmacarthurfoundation.org/epr: https://plastics.ellenmacarthurfoundation.org/epr

EPR Canada. (2017). *Extended Producer Responsibility Summary Report.* EPR Canada. Retrieved January 2022, from http://www.eprcanada.ca/reports/2016/EPR-Report-Card-2016.pdf

Euronews. (2021, July 24). *The Tokyo Olympics will merge ancient heritage with modern innovation.* Retrieved from euronews.com: https://www.euronews.com/green/2021/07/24/the-tokyo-olympics-will-merge-ancient-heritage-with-modern-innovation

European Commission. (2021, July 29). *Circular Economy action plan.* Retrieved from ec.europa.eu: https://ec.europa.eu/environment/strategy/circular-economy-action-plan_en

Eurpoean Commission. (2021, July 28). *A European Green New Deal.* Retrieved from ec.europa.eu: https://ec.europa.eu/info/strategy/priorities-2019-2024/european-green-deal_en#timeline

EXPRA. (2016). *Extended Producer Responsibility.* Extended Producer Responsibility Alliance. Retrieved from https://www.expra.eu/uploads/downloads/EXPRA%20EPR%20Paper_March_2016.pdf

Fothergill, J. (2004). *Scrap Mining: An Overview of Metal Recycling in Canada.* Ottawa: Canary Research Institute for Mining, Environment and Health. Retrieved from https://www.canaryinstitute.ca/publications/Scrap_Mining.pdf

Geyer, R., Jambeck, J. and Law, K. L. (2017, July 19). Production, use, and fate of all plastics ever made. *Science Advances, 3*(7). doi:10.1126/sciadv.1700782

Government of Canada. (2020, October 7). *Canada one-step closer to zero plastic waste by 2030*. Retrieved from Environment and Climate Change Canada: https://www.canada.ca/en/environment-climate-change/news/2020/10/canada-one-step-closer-to-zero-plastic-waste-by-2030.html

Hopewell et al., e. a. (2009, July 27). Plastics recycling: challenges and opportunities. *Philos Trans R Soc Lond. B Biol. Sci.* 364(1526): 2115–2126. doi:10.1098/rstb.2008.0311

IKEA. 2020. *IKEA Sustainability Strategy: People & Planet Positive*. Retrieved July 2021, from https://gbl-sc9u2-prd-cdn.azureedge.net/-/media/aboutikea/pdfs/people-and-planet-sustainability-strategy/people-and-planet-positive-ikea-sustainability-strategy-august-2020.pdf?rev=3a3e9a12744b4705b9d1aa8be3b36197&hash=099EADD58A6B850BD522866B8E01F518

INNOWO. 2020. *How do effective deposit refund systems work?* Innovation Norway. Retrieved from https://innowo.org/userfiles/deposit%20refund%20systems%20Manual%20ENG.pdf

Interface Inc. 2021. *About Net-Works*. Retrieved from net-works.com: https://net-works.com/about-net-works/#next-steps

ISO. (2018). *ISO/TC 323 Circular Economy*. Retrieved from ISO Technical Committees: https://www.iso.org/committee/7203984.html

ISO 20400. (2017). *Sustainable Procurement*. Geneva: International Organization for Standardization.

McDonough, W. 2000. A Boat for Thoreau. *The ruffin Series of the Society for Business Ethics*, pp. 115–133. doi:10.5840/ruffinx2000210

Net-works. (2021, July). *Net-wroks*. Retrieved from net-works.com: https://net-works.com/

OECD. 2014. The State of Play on Extended Producer Responsibility (EPR): Opportunities and Challenges. *Global Forum on Environment: Promoting Sustainable Materials Management through Extended Producer Responsibility (EPR)*. Tokyo: OECD. Retrieved Jan 12, 2022, from https://www.oecd.org/environment/waste/Global%20Forum%20Tokyo%20Issues%20Paper%2030-5-2014.pdf

Ontario Ministry of Environment, Conservation and Parks. (2021, June 3). *Ontario Enhancing Blue Box Program*. Retrieved from https://news.ontario.ca/: https://news.ontario.ca/en/release/1000259/ontario-enhancing-blue-box-program

Park Market and Refillery. (2022, January). *How it Works*. Retrieved from parkmarketandrefillery.com: https://parkmarketandrefillery.com/pages/how-it-works

PlasticBank. (2021, July). *Plastic Bank is empowering the regenerative society*. Retrieved from plasticbank.com: https://plasticbank.com/about/

Plasticoceans.org. 2021. *Plastic Oceans: The-Facts*. Retrieved from Plasticoceans.org: https://plasticoceans.org/the-facts/

PlasticsEurope. 2020. *Plastics—the Facts 2020*. PlasticsEurope. Retrieved July 29, 2021, from https://issuu.com/plasticseuropeebook/docs/plastics_the_facts-web-dec2020

Quebec Ministere de l'Environment. (2021). *Modernization of Québec's deposit and selective collection systems*. Retrieved from www.environnement.gouv.qc.ca: https://www.environnement.gouv.qc.ca/matieres/consigne-collecte/index-en.htm

RCO. (2019, April 29). *Canada recycles just 9 per cent of its plastics*. Retrieved from Recycling Council of Ontario: https://rco.on.ca/canada-recycles-just-9-per-cent-of-its-plastics/

Refillery Market. (2021, June). *Refillery Delivery*. Retrieved from refillerymarket.ca: https://refillerymarket.ca/pages/delivery/,

Roser, Ritchie. 2018. *Plastic Pollution*. Retrieved from Our World in Data: https://ourworldindata.org/plastic-pollution

Sciencehistory.org. 2022. *History and Future of Plastics*. Retrieved from Sciencehistor.org: https://www.sciencehistory.org/the-history-and-future-of-plastics

Sellito, F. M. (2018, August). Industrial Symbiosis: A Case Study Involving a Steelmaking, a Cement Manufacturing, and a Zinc Smelting Plant. *Chemical Engineering Transactions*, pp. 211–216. doi:10.3303/CET1870036

Statista. 2021. *Annual production of plastics worldwide from 1950 to 2020(in million metric tons)*. Retrieved from Statista: https://www.statista.com/statistics/282732/global-production-of-plastics-since-1950/

Southwest The Magazine, July 2015 Issue, pg 16, https://issuu.com/southwestmag/docs/07_july_15, Original quote attributed to Colleen Barrett.

Terra20. (2021, July). *terra20.com*. Retrieved from terra20.com: https://terra20.com/pages/about-us

The Glass Jar Refillery. (2022, January). *THE GLASS JAR REFILLERY*. Retrieved from https://glassjarrefillery.ca/: https://glassjarrefillery.ca/

UNEP. 2018. *Our planet is drowning in plastic pollution—it's time for change!* Retrieved from www.unep.org: https://www.unep.org/interactive/beat-plastic-pollution/.

USAD. (2022, January). *Uzstatos Sistemos Administratorius*. Retrieved from Grazinti verta: https://grazintiverta.lt/en/about/69.

Watkins, E. e. 2019. *Policy approaches to incentivise sustainable plastic design*. Paris: OECD Environment Working Papers, No. 149, OECD Publishing. doi:0.1787/233ac351.

WWF. 2020. *How to Implement Extended Producer Responsibility (Epr): A Briefing For Governments And Businesses*. World Wildlife Fund. Retrieved from https://wwfint.awsassets.panda.org/downloads/how_to_implement_epr___briefing_for_government_and_business.pdf.

WWF Australia (2021, July 2). The Lifecycle of Plastics. Retrieved from WWF.org.au: https://www.wwf.org.au/news/blogs/the-lifecycle-of-plastics#gs.77pikn.

Chapter 5

Seeing Is Believing

Educational and Volunteer Programming to Address Plastic Pollution in Hamilton Harbour

Steven Watts,[1,*] *Christine Bowen*[1,*] and *Chris McLaughlin*[1,2,#]

INTRODUCTION

Plastic pollution is ubiquitous across the Laurentian Great Lakes (Earn et al., 2021). Plastics polluting aquatic environments originate primarily as litter on land where streams, rivers, and infrastructure then act as downstream conduits (Moore, 2008), providing plastic pathways toward places such as Hamilton Harbour at the far west end of Lake Ontario (Fig. 1). Cable et al. (2017) state that a better understanding of this process is critical to understanding "plastic litter budgets" and underpins the success of future management strategies to reduce plastic litter in surface water bodies and avoid the negative consequences of that waste on aquatic environments.

Cable et al. (2017) tracked plastic pollution from anthropogenic sources to environmental sinks in various locations across the Great Lakes, concluding that "concentrations were highest near populated urban areas and their water infrastructure" (Eriksen et al., 2013; Zbyszewski et al., 2014; Driedger et al., 2015; Grbić et al., 2020). This finding has shown significant relevance to Hamilton Harbour, given its immediately adjacent concentration of population, industry, and urban drainage by numerous tributaries, wastewater, and stormwater infrastructure.

Unfortunately, Henderson and Green (2020) suggest that poor public understanding of the relationship between the personal usage of plastics and the growing problem of plastic waste reaching surface waters presents a significant challenge to reducing the plastics problem. However, there is evidence that a variety of community engagement activities and programs can improve public understanding and consequentially create opportunities to advance behavioural and policy changes (for example, Duckett and Repaci, 2015; Poletti and Landberg, 2021; Soares et al., 2021).

[1] Bay Area Restoration Council, 47 Discovery Drive, Hamilton, Ontario, Canada L8L 8K4.
[2] School of Earth, Environment & Society, McMaster University, Hamilton, Ontario, Canada L8S 4K1.
* Principal authorship is shared.
[#] Corresponding author's: chris@bayarearestoration

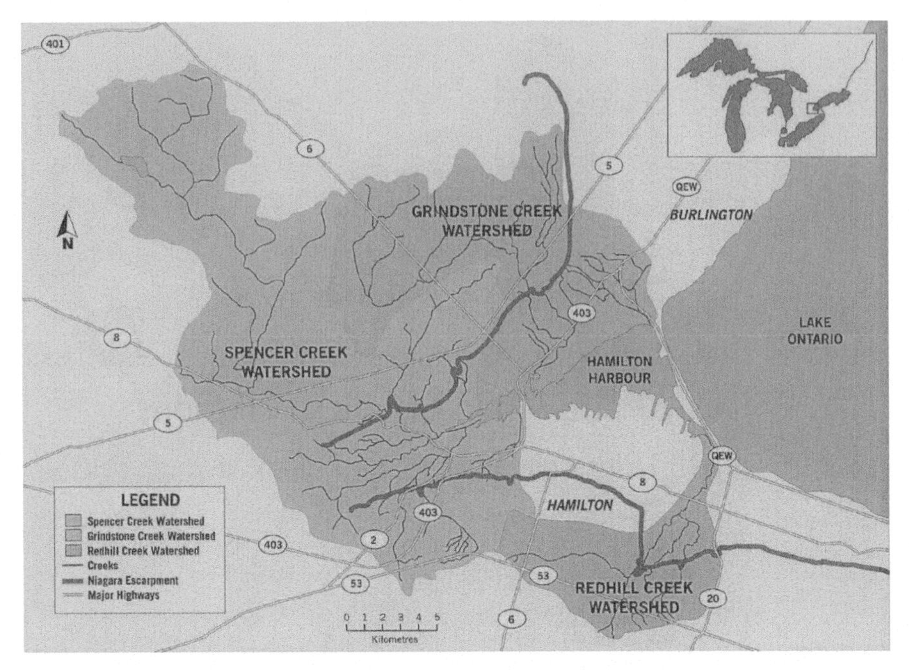

Fig. 1. The Hamilton Harbour Watershed With Its Three Major Subwatersheds in the Context of the Great Lakes. Image property of the Bay Area Restoration Council.

Hamilton Harbour

Hamilton Harbour has been a major shipping port for more than a century. Its southern shoreline is dominated by steel and iron production and is one of the largest concentrations of heavy industry in Canada. Decades of unmitigated industrial waste, untreated wastewater, and urban stormwater severely degraded the Harbour's water quality and ecological health.

In 1985, the International Joint Commission declared Hamilton Harbour as a Great Lakes Area of Concern (AOC) primarily due to its legacy and ongoing pollution that had overwhelmed water quality, eliminated habitat, and contaminated fish and wildlife populations. Those problems of pollution and degradation were labelled beneficial use impairments or BUIs. Remedial Action Plans, or RAPs, were created by government agencies, community organizations, and interested individuals together, a collective process to develop and implement actions that would remove the impairments and restore the beneficial use.[1]

[1] For the most current information on Great Lakes Areas of Concern, corresponding Remedial Action Plans and the status of Beneficial Use Impairments, refer to the websites of the International Joint Commission (ijc.org), Environment and Climate Change Canada (canada.ca/en/environment-climate-change.html), and the U.S. Environmental Protection Agency (epa.gov). For an overview of the strengths and limitations of that collective process relevant to community engagement, see McLaughlin and Krantzberg (2018, 2021).

Plastic pollution was not identified as a specific problem, although one of the 14 categories of BUIs is the Degradation of Aesthetics. The Aesthetics BUI addresses how the water looks and smells as a result of substances that produce persistent objectionable deposits or unnatural colour, cloudiness, or odour. Materials and substances that degrade aesthetics include oil slicks, surface scum, combined sewer overflows, excessive dust, nuisance algal blooms, and debris, such as micro- and macro-plastic waste (i.e., plastic particles < 5 mm and > 5 mm).

The aesthetics of water is a highly subjective measure of environmental quality based on individual perceptions. Although aesthetics can be difficult to quantify and perceptions challenging to track, there was no doubt in 1969 that aesthetics in Hamilton Harbour had been utterly ruined when the member of Parliament for Hamilton-Wentworth, Colin Gibson, described the harbour as a "stinking, rotten quagmire of filth and poisonous waste" (Library of Parliament, 1969). Not surprisingly, the community surrounding the harbour and the city of Hamilton, in particular, had become arguably the most estranged from its water anywhere in Canada.

As with communities across the Great Lakes at the time, community members were among the first to advocate for a change in these dreadful conditions. In Hamilton, the first such community group and the precursor to the Bay Area Restoration Council (BARC) was organized in 1971. As government agencies and other organizations began to develop the Hamilton Harbour Remedial Action Plan (HHRAP) in the late 1980s, many of those same individuals became members of the HHRAP's public advisory committee. As implementation of the HHRAP began, those founding organizers and advocates incorporated BARC in 1991 to represent the public interest in the HHRAP process, to help its fellow community members to recognize and appreciate projects and challenges involved in the restoration, and to monitor and report on the HHRAP's implementation and progress.

BARC has served as an outreach function for 30 years with school programming and volunteer activities and by hosting community engagement events, both in-person and increasingly online events. These varied approaches are a tremendous opportunity to shape public perceptions of the challenges and successes of restorative actions. In 2019, for example, BARC had more than 17,000 participants in its school programs and events that engaged with varied water literacy content and activities. In the last several years, schools and volunteer programming have increasingly included community science activities with the participants. Through this engagement, BARC staff has detected an increasing community interest and concern over plastic pollution.

The growing public interest in plastic pollution is being expressed to BARC with new programming that educates students and the public about the complex connections of people and their plastic waste to the Harbour via tributaries and infrastructure. In addition, BARC has been contracted by federal and provincial agencies in recent years to engage substantively in the Aesthetics BUI. The subjective nature of this BUI means measuring progress in reducing the impairment of aesthetic conditions and determining the status of the BUI which is a significant and unresolved challenge.

This chapter shares BARC's varied approaches to engaging and educating the community about plastic pollution in local aquatic environments. We describe the public understanding of plastic pollution to the extent that BARC staff have engaged with students, volunteers, and others to understand, quantify, and reduce plastic pollution in the Harbour. We have approached this chapter as educators and practitioners within the HHRAP community to highlight the importance of bridging scientific knowledge and public understanding to encourage and assist others with similar missions for engagement and education. The delivery of impactful programming is designed to educate and provide opportunities to directly engage community members in solutions which improve public understanding of plastic pollution and take advantage of local enthusiasm for problem-solving with tangible actions and measurable outcomes.

Educational and Volunteer Programming

Creating volunteer opportunities for community members and delivering educational programming for youth are key aspects of BARC's mandate. Opportunities are designed to provide substantive experiential learning for participants, and the scope and content of each program is tailored to align with the provincial school curriculum at varying grade levels. BARC delivers four activities or programs that engage participants in issues of plastic pollution: litter cleanups, Seabin Project events, Water School, and shoreline monitoring. The delivery methods of BARC's programs are an important aspect of the success of our community engagement. Specific approaches are used to reach broad demographics and ensure each participant gains the intended information and experience.

Falk's visitor identity model

The demographics, backgrounds, and diversity of the BARC program's participants vary significantly, and participant motivations are assessed; these are considered in the development and delivery. Falk's (2009) research on free-choice learning, museum visitor studies, and science education informs BARC's approach when considering different participant motivations. Falk's predictive model is valuable in BARC's attempts to best meet the needs of program participants. Although we cannot always predict participant motivation prior to events, Falk's model is useful for preparing content and activities that engage specific audiences or varied participants. Falk outlines five identities: *explorers*, *facilitators*, *professionals/hobbyists*, *experience seekers*, and *rechargers*. BARC encounters each visitor's identity throughout the delivery of various outreach programs.

Explorers are motivated by curiosity and an interest in learning about topics, engaging with content that helps them learn, and gaining an understanding of issues. In Falk's study, explorers described themselves as curious, science lovers, learners, and discoverers (also see Cotter et al., 2021).

Facilitators are guided by social motivations and view event participation as an opportunity to spend time with those who accompany them (for example, a parent bringing a child). Participation by facilitators does not necessarily lead to impactful education as it sometimes becomes more about the opportunity to socialize. Their intention is to satisfy the needs or desires of the people they bring with them.

Professionals/hobbyists are driven by their specific interests and choose events that build on their background or existing knowledge. In the case of plastic pollution, for example, professionals/hobbyists likely already have an interest in environmental issues and attend BARC events to gain specific and related knowledge and experience. They often come with a goal in mind and are described as being 'on a mission' to achieve it.

Experience seekers are motivated by a desire to collect experiences and attend events for recreational opportunities. They expect to have an enjoyable experience and have the desire to leave fulfilled.

Rechargers are looking for events or information that is introspective and an opportunity to rejuvenate. When motivated to recharge, these participants hope to experience awe during the event, such as seeing natural beauty (Cotter et al., 2021).

Tilden's principles of heritage interpretation

The development and delivery of BARC's programs also use Tilden's (1977) six principles of heritage interpretation—described as "the communication of information about, or the explanation of, the nature, origin, and purpose of historical, natural, or cultural resources, objects, sites, and phenomena using personal or non-personal methods"—as a way to convey stories, information, and messages in ways that are approachable and digestible for broad demographics. BARC's programs aim to reveal relatable and holistic information in an inclusive and age-appropriate manner as a way to inspire emotionally-driven (and knowledge-based) action. Incorporating Tilden's principles helps to ensure that educational and volunteer programming is impactful, relatable, and directed to a varied audience. The six principles of heritage interpretation are as follows (Tilden, 1977):

1. Interpretation that does not somehow relate what is being displayed or described to something within the personality or experience of the visitor will be sterile. Interpretation should be personal to the audience.
2. Information, as such, is not interpretation. Interpretation is revelation based on information. Successful interpretation must do more than just present facts.
3. Interpretation is an art which combines many forms of art. Any art is, to some degree, teachable.
4. The chief aim of interpretation is not instruction but provocation. Interpretation should stimulate people toward a form of action.
5. Interpretation should aim to present the whole rather than the part. Interpretation is conceptual and should explain the relationships between things.
6. Interpretation addressed to children should not be a dilution of the presentation to adults, but it should follow a fundamentally different approach. Different age groups have different needs and require different interpretive programs.

BARC's volunteer and education programs

This section provides a brief overview of four examples of BARC programs or events that both educate and engage students and other community members with varied objectives and intended outcomes. Each example references Falk's model and Tilden's principles and describes BARC staff's experience with the development and delivery of the programs and events.

Litter cleanups (facilitators, explorers, and rechargers)

Litter cleanup events involve the planning and delivery of an event in which volunteers remove litter and debris from a shoreline, recreational, or natural area. The shorelines of Hamilton Harbour are well-used by the general public for recreation and pleasure. Due to regular usage and high pedestrian and recreational traffic, significant litter accumulates in these areas. In this chapter, we will not be analyzing the motivations of people who litter, rather we are focussing on the motivations of individuals who attend cleanup events.

The litter, found along shorelines, is made up of different materials, including various forms of plastic. Once discarded, plastic litter breaks down into small pieces, resulting in microplastics (MPs). These MPs eventually get into waterways, causing issues for wildlife, public health, and the aesthetics of the Harbour.

Volunteers at cleanup events are often facilitators, explorers, or rechargers based on Falk's predictive model. Based on volunteer history, attendees include families, community groups, and those who have prior volunteer experience with BARC. Cleanup events are always located in the natural environment which provides an opportunity for volunteers to appreciate and learn about local ecosystems.

Cleanup events provide the tools and chosen locations for removing debris. The tools provided, such as garbage bags and gloves, are only one part of litter cleanup. In addition to the manual removal of debris, BARC also provides relevant information about the impacts of litter on the overall health of Hamilton Harbour. Participants in these events are motivated to not only clean up litter but also want to learn about the impact of pollution. The resulting pile of garbage bags filled with the litter picked up by our volunteers is a visual indicator of that impact. Following Tilden's second principle, the pile of filled garbage bags clearly reveals the issue to participants. Seeing the litter they have removed is a signal of the volume of litter discarded near our waterways. The sight of filled garbage bags also provides provocation for changed behaviour which corresponds accordingly to Tilden's fourth principle.

Participants who attend cleanups initially look for large pieces of litter, such as single-use coffee cups or water bottles. BARC informs volunteers about the presence of smaller microplastics (i.e., "MPs" or plastics < 5 mm) and the importance of removing those items. Volunteers at cleanup events are informed that common single-use items such as cigarette filters, for example, are made up of plastic fibers that are shed when cigarettes are littered into waterways, and that leachate from cigarette butts is acutely toxic to marine and freshwater fish species. That research, by Slaughter et al. (2011), demonstrated that leachates from smoked cigarette butts with remnant tobacco were significantly more toxic to fish than the smoked filters alone, but that unsmoked filters also exhibited a small level of toxicity.

By providing participants with additional information about the anthropogenic impacts of plastic pollution, especially on recreational and livelihood activities like fishing, Tilden's first, fourth, and fifth principles are met. BARC is able to build a connection, draw an emotional response, and present a holistic picture for participants. In doing so, BARC also exercises two foundations of environmental education: highlighting our interdependence with the natural world and integrating multiple disciplines, such as natural sciences and social sciences, into program content (NAAEE, 2010).

Seabin project events (professionals/hobbyists and explorers)

The Seabin is a floating garbage receptacle that works by suctioning surface water and all debris at or near the surface into a catch bag and then retaining any litter and MPs larger than 2 mm.[2] There is evidence that these types of interventions in plastic waste pathways across watersheds can both attract public attention to the plastic problem (McNeil, 2019) and be influential in problem-solving (Smyth et al., 2021). In Hamilton Harbour, the Hamilton-Oshawa Port Authority (HOPA) installed three Seabins in its Harbour West Marina in the summer of 2021 when BARC staff and volunteers began working with HOPA marina staff to characterize (i.e., sort, describe, and quantify) the waste collected. BARC has partnered with HOPA to deliver these waste characterization events as participating local agencies in the Great Lakes Plastic Cleanup (GLPC) initiative that has focused on removing plastics from these nearshore areas right across the Great Lakes.[3] The GLPC is a first-of-its-kind initiative that uses this innovative technology to analyze and communicate the problem of plastic pollution in the Great Lakes.

The characterization of waste in Seabins allows BARC to identify what waste is most prevalent in the marina and then dispose of it appropriately. Volunteers sort the waste found in the Seabins by size and type. On the worst days, the Seabins in Hamilton Harbour collected 2 kg of waste per day. Preliminary results suggest cigarette butts and small pieces of plastic film are the most frequent sources of local plastic pollution.

Seabin waste characterization events attract professionals/hobbyists and explorers. Volunteers for this program usually have background knowledge of water issues in Hamilton. The Seabin events attract fewer participants than general cleanups and involve an analytical/data collection aspect. Professionals/hobbyists seek to gain experience with MP data collection and to learn about the impacts of plastic pollution in the Harbour. Explorers are curious about the type of litter found in the marina and learn about their reduction strategies. In an effort to build connections, BARC finds it beneficial for participants to share their existing knowledge and experience. In order to allow revelations to occur, participants next brainstorm the intention of the program and speculate in advance on what plastics might have been collected in the

[2] The Seabin Project works with marinas, ports, and yacht clubs to reduce marine litter with an initiative that has evolved into a comprehensive research, technology, and educational initiative with global interest and reach. For more information about the Seabin Project, visit seabinproject.com.

[3] For more information about the Great Lakes Plastic Cleanup, visit greatlakesplasticcleanup.org and see the chapter by Hilkene et al. in this volume.

Seabins. Much like the general cleanup events, participants are also informed about MPs and the overall health of Hamilton Harbour. As many professionals/hobbyists are in attendance, we specifically reference any relevant academic papers and direct participants to more local resources from our partners. A first-hand connection with the Harbour, combined with a sharing of personal connections and history, creates an opportunity for place-based education (Smith, 2002).

Water school (facilitators and explorers)

Hamilton Harbour is fed by numerous streams and creeks that bisect neighbourhoods across the city on the surface and below (buried in stormwater infrastructure). These waterways and catchments include the properties of schools that participate in BARC's many classroom and playground programs and events. However, school communities do not often have strong relationships with the water flowing near their school. Students lack first-hand experiences with their local environment which is a necessity for effective environmental education (Smith, 2002). The school community may have an awareness that nearby waterways exist, but in our experience do not usually know their name, health, or physical characteristics.

Environmental degradation, which is combined with an inability for people to recognize those degraded conditions, can lead to a negative feedback cycle in which the decline in ecosystem health occurs without notice or care. To break this cycle, attention needs to be placed on creating new connections between people and nature and strengthening those connections where they are weak (Schuttler et al., 2019). As such, BARC created Water School—a place-based program designed to connect secondary school communities to their immediate local waterways—so that student participants can see, feel, and build a relationship with those waterways. For environmental education to be most effective, it must concentrate on connections to environments where people live (NAAEE, 2010). Water School provides a multi-day immersive experience of water testing by allowing and requiring students to engage directly with flowing water. Students learn about the extent and characteristics of the Harbour's watershed, conduct experiments and surveys, and become familiar with water testing equipment in the field. This program is rooted in practical and hands-on experience and provides opportunities for investigation and analysis which is a foundational principle of environmental education (NAAEE, 2010).

Water School engages students and their teachers in a program with classroom activities, field experience, and reflections. Teachers, often identified as facilitators, reach out to BARC to schedule this program and identify the specific needs of their students. Prior to program delivery, BARC's planning involves conducting site research at local streams and forming strategies to make the program most effective and enjoyable. Teachers also work with BARC to connect Water School activities to the provincial education curriculum as well as any ongoing student projects and interests. Water School focuses on chemistry; however, there is an opportunity to meet broader curriculum ties and make the activities cross-disciplinary, for example by using math via run-off calculations or creating art using litter collected. Students, who become curiously engaged in water chemistry during the program, are identified as explorers under Falk's model. While conducting water testing experiments, students begin to physically explore and engage with the stream. Specifically, they undertake an aesthetics survey and use all of their senses to evaluate the stream using specific criteria.

Water School provides the tools for students and teachers to understand and recognize water quality and its other properties and characteristics in their neighbourhood. This process involves developing a relationship between BARC and the school community. BARC provides skill, information, and equipment to lead experimental, investigative activities and in the process, BARC staff become teachers and guides while remaining relatable to the young audiences. Students value a sense of personal and professional relatability during programming. It provides a sense of trust and a deeper investment in the learning process. To do so, BARC includes scientific facts about plastic pollution that relates or applies to the lived experiences of students. BARC staff also explain processes that need to be followed during experimentation. They explain that there are reasons why students are testing for specific parameters which makes the students then curious to know the purpose of each test or measurement. It is important to have open discussions with students and incorporate different forms of learning and activities. Although significant planning goes into the program, BARC cannot predict the exact learning style and experiences of each student which is why diversity of information and activities can improve the chances of program success. We are inclusive of all participants as the connections students develop are part of a personal process. Furthermore, by providing a variety of information and activities and taking a systems approach to environmental education, BARC appeals to Tilden's principle of a holistic approach to engagement.

Moreover, BARC staff forge a personal connection with students to establish a strong working relationship. Building this relationship often requires sharing personal experiences and opinions about topics, issues, and problems. Students appreciate hearing career advice and how staff have learned from their mistakes. Throughout the program, this relationship is strengthened as students become comfortable asking questions and engaging in experiments.

Shoreline monitoring (professionals/hobbyists, explorers)

Like Hamilton Harbour, Toronto Harbour, which is further east along the north shore of Lake Ontario, has also suffered historical pollution problems from inadequate treatment of industrial wastewater, sewage discharges, excessive floating debris, odour, and unnatural turbidity. These issues were captured by the Aesthetics BUI in the Toronto and Region Remedial Action Plan (TRRAP), and Dahmer et al. (2018) described the qualitative monitoring program implemented there to measure and assess aesthetic conditions in Toronto Harbour and parts of its watershed. An Aesthetic Quality Index (AQI), developed elsewhere for use in Areas of Concern to measure progress toward removing the problems impairing aesthetics, was adapted for the TTRAP by taking advantage of existing monitoring programs and local expertise.

BARC was contracted in 2018 by senior government agencies of the HHRAP to report on the status of the Aesthetics BUI in Hamilton Harbour; in part, it compiled aesthetics data collected between 2012 and 2017 and attempted a semi-quantitative assessment using the TTRAP AQI assessment methodology to conduct further investigations in 2018 and to augment those data, including research, surveys, and fieldwork. The report (BARC, 2019) was to make recommendations and provide direction regarding the adequacy of evidence, plans, and programming to address the status of the Aesthetics BUI for the HHRAP and to ensure continued and improved aesthetic conditions were achieved and monitored in the future.

At the time of this writing, the final methodology to assess the status of aesthetics in Hamilton Harbour has not been agreed upon by a consensus of HHRAP agencies. However, BARC continues to develop a program for community science participants—specifically, the trained young adult volunteers who make substantive commitments of skill and time to our Community Water Leaders program—to collect data and monitor and report to the public on aesthetic conditions in the Harbour and across its watershed. Like the Seabin project, volunteers have background knowledge of local water issues and the programs attract the fewest number of participants compared to general cleanups due to the skill and time required. The shoreline monitoring attracts professionals/hobbyists volunteers due to the analytical nature of the tasks, and these volunteers seek to gain experience with plastic pollution data collection and to learn about its impacts on aquatic environments. Explorers are also amenable to these tasks as they are curious about the type of litter found along shorelines and their enthusiasm for learning also makes them able to be trained for the type of semi-quantifiable data collection involved. But the focus is on attracting professionals/hobbyists for participation due to the diligence and rigour of the tasks. BARC continues to work toward an integration of governmental and nongovernmental agency support and cooperation on resolving issues of logistics, methodology, and public communications of the Aesthetics BUI specifically and environmental quality at large.

Discussion

Prior to the Hamilton Harbour RAP, industrial land uses and development had almost eliminated public access at the Harbour's southern and western shorelines and, as described, water quality and aesthetic conditions were utterly dreadful. Today, however, a walk around Bayfront Park and along the Waterfront Trail can be a full sensory experience; many pleasant sights, sounds, and smells have returned, and a physical connection to the water is possible. These ongoing changes have been made possible and reinforced by efforts to engage the broader community with programs and activities that provide teaching opportunities and tangible experiences for the general public. The most popular public programs typically involve issues that are readily seen and sometimes readily smelled. Issues like algae blooms and sewage bypasses are clear and recognizable, and they can impact public uses and perceptions of the Harbour. However, with issues that cannot be readily seen, such as the sources, extent, and impacts of MPs, there need to be more directed efforts to educate and engage the community.

BARC's educational and volunteer programming is an important foundation for engaging the Harbour's watershed community in hidden problems, such as MPs. Through purposeful action, such as school programs, shoreline cleanup events, and community science initiatives, there is an opportunity to encourage and nurture an engaged public for positive outcomes. Such opportunities require a network of stakeholders, community partners, and individuals to adopt a shared perception of MPs, to collectively "see" MPs, and work toward community action for environmental benefit.

A shared perception of MPs, however, first requires a collective understanding of the issue. The scientific research behind MPs is vast and growing and is most often shared in publications and venues that are beyond the access of the general public. One of the challenges to achieving a shared understanding is putting that science and understanding into the hands and minds of those community members most likely to use it and respond to new knowledge. Educational and outreach programs like the ones offered by BARC serve to connect scientists and other professionals with those public audiences and improve water and environmental literacy in the broader community by seeking to connect with the varied motivations for people to become involved (see Duckett and Repaci, 2015). BARC works to provide and translate scientific research on MPs, for example, for a range of demographics and backgrounds because "seeing" MPs is a critical element for further understanding, appreciation, and advocacy on a problem that will require multiple avenues to solve.

References

Bay Area Restoration Council. 2019. Status Report on the Degradation of Aesthetics Beneficial Use Impairment XI in the Hamilton Harbour Remedial Action Plan. https://hamiltonharbour.ca/resources/documents/BARC_2019_Degradation_of_Aesthetics_BUI_Status_Report.pdf [accessed August 25, 2021].

Cable, R.N., Beletsky, D., Beletsky, R., Wigginton, K., Locke, B.W. and Duhaime, M.B. 2017. Distribution and modeled transport of plastic pollution in the Great Lakes, the world's largest freshwater resource. *Front. Environ. Sci.* 5: Art. 45.

Cotter, K., Fekete, A. and Silvia, P. 2021. Why do people visit art museums? Examining visitor motivations and visit outcomes. *Empir. Stud. Arts* (in press).

Dahmer, S.C., Matos, L. and Jarvie, S. 2018. Assessment of the degradation of aesthetics Beneficial Use Impairment in the Toronto and region Area of Concern. *Aquat. Ecosyst. Health Manag.* 21(3): 276–284.

Driedger, A.G.J., Dürr, H.H., Mitchell, K. and Van Cappellen, P. 2015. Plastic debris in the Laurentian Great Lakes: *A review. J. Great Lakes Res.* 41: 9–19.

Duckett, P. and Repaci, V. 2015. Marine plastic pollution: using community science to address a global problem. *Mar. Freshw. Res.* 66: 665–673.

Earn, A., Bucci, K. and Rochman, C.M. 2021. A systematic review of the literature on plastic pollution in the Laurentian Great Lakes and its effects on freshwater biota. *J. Great Lakes Res.* 47: 120–133.

Eriksen, M., Mason, S., Wilson, S., Box, C., Zellers, A., Edwards, W., Farley, H. and Amato, S. 2013. Microplastic pollution in the surface waters of the Laurentian Great Lakes. *Mar. Pollut. Bull.* 77: 177–182.

Falk, J.H. 2009. Identity and the Museum Visitor Experience. Routledge, New York.

Grbić, J., Helm, P., Athey, S. and Rochman, C.M. 2020. Microplastics entering northwestern Lake Ontario are diverse and linked to urban sources. *Water Res.* 174: 115623.

Henderson, L. and Green, C. 2020. Making sense of microplastics? Public understandings of plastic pollution. *Mar. Pollut. Bull.* 152: 110908.

Library of Parliament, 1969. House of Commons Debates. First Session, Twenty-Eighth Parliament. Volume IX, Page 9262. https://parl.canadiana.ca/view/oop.debates_HOC2801_09/116?r=0&s=1 [accessed September 15, 2018].

McLaughlin, C. and Krantzberg, G. 2018. Remedies for improving Great Lakes Remedial Action Plans: a policy Delphi study. *Aquat. Ecosyst. Health Manag.* 21: 493–505.

McLaughlin, C. and Krantzberg, G. 2021. Learning from experience with Great Lakes Remedial Action Plans: A policy delphi study of development and implementation. *In*: J.H. Hartig and M. Munawar (eds.). Ecosystem-Based Management of Laurentian Great Lakes Areas of Concern: Three Decades of U.S. - Canadian Cleanup and Recovery. Michigan State University Press, East Lansing, USA.

McNeil, M. 2019. Fishing for garbage in Hamilton Harbour. Hamilton Spectator, September 7, https://www.thespec.com/news/hamilton-region/2019/09/07/fishing-for-garbage-in-hamilton-harbour.html [accessed August 20, 2021].

Moore, C.J. 2008. Synthetic polymers in the marine environment: a rapidly increasing long-term threat. *Environ. Res.* 108: 131–139.

Poletti, S.A. and Landberg, T. 2021. Using nature preserve creek cleanups to quantify anthropogenic litter accumulation in an urban watershed. *Freshw. Sci.* 40: 537–550.

Schuttler, S. et al. 2019. Citizen science in school: students collect valuable mammal data for science, conservation, and community engagement. *BioScience* 69: 1.

Slaughter, E., Gersberg, R.M., Watanabe, K., Rudolph, J., Stransky, C. and Novotny, T.E. 2011. Toxicity of cigarette butts, and their chemical components, to marine and freshwater fish. *Tob. Control.* 20(Suppl 1): i25–i29.

Smith, G. 2002. Place-based education: learning to be where we are. *Phi Delta Kappan* 83: 584–594.

Smyth, K., Drake, J., Li, Y., Rochman, C., Van Seters, T. and Passeport, E. 2021. Bioretention cells remove microplastics from urban stormwater. *Water Research* 191, p. 116785.

Soares, J., Miguel, I., Venâncio, C., Lopes and Oliveira, M. 2021. Public views on plastic pollution: Knowledge, perceived impacts, and pro-environmental behaviours. *J. Hazard. Mater.* 412: 125227.

Tilden, F. 1977. Interpreting our heritage. 3rd edition. University of North Carolina Press, USA.

Zbyszewski, M., Corcoran, P.L. and Hockin, A. 2014. Comparison of the distribution and degradation of plastic debris along shorelines of the Great Lakes, North America. *J. Great Lakes Res.* 40: 288–299.

Chapter 6

The Great Lakes Plastic Cleanup

An Effective Approach to Addressing Plastic Pollution in the Great Lakes

Melissa DeYoung,[1,*] *Mark Fisher*[2] and *Christopher Hilkene*[1]

INTRODUCTION

Plastics are one of the most revolutionary innovations of the modern world. They play a critical role in shaping our economy and our daily lives. The functionality, durability, low maintenance, and relative cost-effectiveness of plastic products have contributed to significant growth in their application since the early 1950s. Today, plastics are ubiquitous with over a thousand types available in the market, and their versatility makes them appealing for a range of applications, including child safety helmets, airbags, cell phones, computers, packaging applications, and other electronic equipment. However, the way in which plastic is used, managed, and disposed of can pose significant challenges to the environment and our health.

While many plastics have the ability to be recovered, recycled, or reconstituted, this value is lost when they are discarded or improperly disposed of. Improper disposal also contributes to extracting natural resources to create new plastics, generating greenhouse gas (GHG) emissions, polluting water systems, and releasing harmful pollutants. The durability of plastic means that most types do not break down easily in the environment, thus creating major challenges around their appropriate disposal.

Each day, plastic debris and fibres make their way into rivers, streams, and lakes where they accumulate. As a result, plastic accumulation, including in the oceans, has become an increasingly visible environmental issue that has led to an incredible global momentum to address plastic pollution in recent years. On 9 June

[1] 130 Queens Quay East, Ste. 902, West Tower, Toronto, Ontario, M5A 0P6, 416-926-1907.
[2] 3247 Clearwater Crescent, Ottawa, Ontario, K1V 7S3, (613) 668-2044.
Emails: mark@councilgreatlakesregion.org; chilkene@pollutionprobe.org
* Corresponding author: mdeyoung@pollutionprobe.org

2018, governments, including those in Canada, France, Germany, Italy, the United Kingdom, and the European Union together with a number of large corporations, endorsed the Ocean Plastics Charter[1]—a voluntary charter outlining concrete actions to eradicate plastic pollution—to demonstrate their commitment to taking concrete and ambitious actions to address the problem. Endorsees commit to moving toward resource efficiency and a sustainable approach to plastics management through action on:

- Sustainable design, production, and after-use markets
- Collection, management, and other systems and infrastructure
- Sustainable lifestyles and education
- Research, innovation, and new technologies
- Coastal and shoreline action

When it comes to plastic waste, the spotlight tends to shine on oceans' plastic which comes with a good reason. However, freshwater systems play a significant role in the story. Plastic waste can have a devastating effect on these critical ecosystems before it eventually makes its way to the ocean and contributes to broader-scale impacts. It is estimated that approximately 10 million kg of plastic flows into the Laurentian Great Lakes alone each year. Microplastics (MPs)—plastic particles < 5 mm in size—have been found in surface water, sediment, and wildlife in and around the lakes with levels in surface waters of some parts of the lakes reported to be as high as 1.25 million particles per km^2; a concentration on par with that was found in the ocean's garbage patches.

The Importance of the Great Lakes

The five Laurentian Great Lakes—Lake Superior, Lake Michigan, Lake Huron, Lake Erie, and Lake Ontario—form the largest group of freshwater lakes on Earth, holding 21% of the Earth and 84% of North America's surface freshwater. A globally significant ecosystem, the lakes are considered one of the most diverse natural systems in the world, supporting 3,500 species of plants and wildlife, including more than 350 species of fish.[2] Some of these species are so singular to the Lakes' ecosystem that they cannot be found anywhere else on Earth.

The Great Lakes border two countries, and within those two countries are eight American states (Illinois, Indiana, Michigan, Minnesota, New York, Ohio, Pennsylvania, and Wisconsin) and one Canadian province (Ontario). A second province, Québec, is also directly affected by Great Lakes water quality. The region is home to hundreds of rural and urban municipalities, starting from Duluth in Minnesota to Montreal in Québec, numerous tribal governments, First Nations, and Métis communities. The Great Lakes (or *Nayaano-nibiimaang Gichigamiin*—an Anishinaabemowin term for the five freshwater seas—as they are referred to by

[1] Government of Canada. 2018. Ocean Plastics Charter. Retrieved from https://www.canada.ca/en/environment-climate-change/services/managing-reducing-waste/international-commitments/ocean-plastics-charter.html.

[2] Governments of Canada & U.S. 2019. 2019 Progress Report of the Parties. Retrieved from https://binational.net/wp-content/uploads/2019/06/Final-2019-PROP-English-June-7.pdf.

many First Nations) also provide drinking water for more than 40 million people, and the region is growing rapidly.

The Great Lakes and surrounding watershed also support a thriving binational economy and cater to more than 1.5 million jobs. The economy rivals that of most nations with an annual GDP of approximately 6.0 trillion USD; if the region is hypothesized as a country, it would be then the third largest economy in the world, behind only the U.S. and China. Each year, approximately 230 million metric tonnes of raw and finished goods, valued at over 100 billion CAD, move through the Great Lakes to St. Lawrence Seaway which is the largest inland deep draft navigation system in the world. The Lakes are a vital shipping route for overseas markets with major ports located along their shores, supporting the global activities of a number of top consumer brands and Fortune 500 companies that are headquartered in the region, such as Procter and Gamble, Dow Inc., Kellogg, General Mills, Archer Daniels Midland, Kraft Heinz, Kohler, Boeing, 3M, Cargill, SC Johnson, Eli Lilly, Whirlpool, John Deer, the Ford Motor Company, and General Motors.

Farming and food production are also vital to the binational economy; according to the U.S. Environmental Protection Agency, nearly 25% of Canadian agricultural production and 7% of American, or roughly 55 million acres, is located within the Great Lakes Basin. Moreover, a third of Canada's employment in agriculture and food processing is found in Ontario, and almost all of this is within the Great Lakes region.[3] In the U.S., the eight Great Lakes states account for roughly 34%, or 8.4 billion CAD, of agriculture and agri-food trade from the U.S. to Canada.

Continued development within the Great Lakes region has had significant impacts on the health of the lakes themselves. While important progress has been made in recent years in protecting their waters and restoring areas that have been severely degraded or contaminated as a result of the region's industrial history, the ongoing development and expansion of the built environment and the growth of the economy continues to affect the watershed, such as in some cases pushing species, habitats, and ecosystems to their limits.

Plastic Pollution in the Great Lakes

Protecting the Great Lakes and preventing future environmental issues is vital to the economic, social, and ecological wealth of not just the region but of the continent as a whole. In recent years, an increasing amount of plastic debris of all types, shapes, and sizes has been found to enter the lakes from a variety of sources (e.g., litter, debris from vessels, tire dust in road runoff, pellets from industry, or microfibers from laundering clothing).

There are many different types of plastic, each of which interacts with the environment in different ways, which has a range of potential impacts on aquatic ecosystems. While data specific to the Great Lakes remains relatively limited, field and laboratory studies suggest that plastic pollution is common in freshwater food webs with the potential for immediate and long-term environmental impacts. For

[3] Province of Ontario 2017. Ontario's Great Lakes Strategy. Retrieved from https://www.ontario.ca/page/ontarios-great-lakes-strategy#section-3.

instance, a recent study found that macroalgae in the Great Lakes contained large quantities of synthetic microfiber pollutants, including polyethylene terephthalate (PET) and polyethylene, both entangled and adsorbed into the algae strands. These microfibers have the potential to find their way into the food web, affecting ecological communities in the lakes.[4]

A recent laboratory study demonstrated the transfer of nano-sized polystyrene in a freshwater food chain, including algae, water fleas, and two levels of fish.[5] Polystyrene has been shown to adsorb other contaminants, such as polychlorinated biphenyls (PCBs) and polycyclic aromatic hydrocarbons (PAHs), which can be passed on to other organisms through ingestion. Potential impacts on freshwater fish in the study included damage to liver tissue, altered metabolism, disturbance to embryos, and changes in locomotive activity.[6]

Polyethylene and polypropylene plastic have also been found in the gastrointestinal tract of double-crested cormorant chicks in the Great Lakes with studies suggesting that this may be due to parent cormorants consuming prey (e.g., fish), which have themselves consumed plastic before being captured, and then feeding this prey to their chicks.[7] Another study also supports this after finding traces of MPs, including fibres, in the digestive tracts of fish in three major tributaries of Lake Michigan.[8]

Fish larvae and invertebrates are also at risk. In laboratory studies, chronic exposure to polyethylene particles and polypropylene microfibers was found to significantly decrease freshwater amphipod growth and reproduction, and exposure to polyvinyl chloride (PVC) significantly inhibited weight gain and growth of common carp larvae. In the Grand River watershed, which flows into Lake Erie, freshwater mussels were found to be ingesting MP fragments, including those made of polypropylene.[9]

Despite mounting evidence of its potentially devastating impacts, plastic pollution in our freshwater ecosystems, including the Great Lakes, continues to be largely overlooked. Bolstered by a growing body of research, Great Lakes organizations and advocates began to sound the alarm bell over a decade ago and to call for more coordinated actions by governments, industries, scientists, and civil society organizations in both the U.S. and Canada to combat the issue. Pollution Probe and the Council of the Great Lakes Region have been at the forefront of efforts to address the plastic waste and pollution issue from a freshwater perspective.

[4] Peller J. et al., 2021. Sequestration of microfibers and other MPs by green algae, and Cladophora in the US Great Lakes. Environmental Pollution, 276. 116695.

[5] Chae Y. et al., 2018. Trophic transfer and individual impact of nano-sized polystyrene in a four-species freshwater food chain. Scientific Reports, 8, 284.

[6] Rochman, C.M., Hoh, E., Kurobe, T. and Teh, S. J. 2013. Ingested plastic transfers hazardous chemicals to fish and induces hepatic stress. Scientific Reports, 3, 3263.

[7] Brookson et al., 2019. MPs in the diet of nestling double-crested cormorants (Phalacrocorax auratus), an obligate piscivore in a freshwater ecosystem. Canadian Journal of Fisheries and Aquatic Sciences 76, 11.

[8] McNeish et al., 2018. MP in riverine fish is connected to species traits. Scientific Reports, 8, 11639.

[9] Xia et al., 2020. Polyvinyl chloride MPs induce growth inhibition and oxidative stress in Cyprinus carpio var. larvae. Science of the Total Environment, 716, 136479.

As part of its G7 presidency, the Government of Canada convened the G7 Ministers of Environment and Energy in Halifax, Canada, in September 2018. The Council of the Great Lakes Region and Pollution Probe were invited to attend the event which focused on climate change, oceans, and clean energy and provided an important opportunity to engage directly with then Minister of Environment and Climate Change Canada, the Honourable Catherine McKenna, and U. S. Administrator of the Environmental Protection Agency, the Honorable Andrew Wheeler, about the growing plastic pollution problem in the Great Lakes.

In 2018, Pollution Probe and the Council of the Great Lakes Region hosted the Great Lakes Plastics Forum in Toronto, Canada. The forum provided an opportunity for a range of experts to convene and exchange ideas about how best to work together to develop innovative and practical solutions to the issue of plastic waste and improper disposal while supported by sound public policy.[10] In 2020, the two organizations were invited by Environment and Climate Change Canada to inform and co-host a workshop in Toronto in support of the development of Phase 2 of the Canadian Council of Ministers of the Environment's Canada-Wide Action Plan to Achieve Zero Plastic Waste.[11] The workshop and subsequent action plan placed significant focus on:

- improving consumer, business, and institutional awareness to prevent and manage plastic waste responsibly
- reducing plastic waste and pollution generated by aquatic activities
- advancing plastics science to inform decision-making and measure performance over time
- addressing plastics in the environment through capture and clean-up
- contributing to global action on plastic pollution reduction.

Building on these efforts, the Council of the Great Lakes Region and Pollution Probe began exploring further opportunities to work collectively with governments, the private sector, academia, and environmental and community groups to create innovative and practical solutions to plastic pollution in the Great Lakes that would have immediate, tangible results. The Great Lakes Plastic Cleanup—an initiative of the Pollution Probe and the Council of the Great Lakes Region and supported by a broad network of collaborators—was launched in 2020. The initiative is the largest of its kind in the world, uniquely combining a focus on removing plastic from the environment, community outreach and engagement, and the collection of locally-relevant data to better inform regulatory and policy decisions. A number of key learnings have emerged from the first two years of the initiative related to the important role, including the combination of innovative technology, data collection, and local engagement which can successfully address an environmental issue of

[10] Pollution Probe and the Council of the Great Lakes Region. 2018. Great Lakes Plastics Forum: Solutions for a Sustainable Future Summary Report. Retrieved from https://www.pollutionprobe.org/wp-content/uploads/Great-Lakes-Plastics-Forum-Summary-Report.pdf.

[11] Canadian Council of Ministries of the Environment. 2020. Canada-Wide Action Plan on Zero Plastic Waste: Phase 2. Retrieved from https://ccme.ca/en/res/ccmephase2actionplan_en-external-secured.pdf.

significance and discuss how it can further contribute to necessary systemic changes required to address plastic pollution in the region.

The Role of Plastic Capture Technologies in Driving Systems Change

Over the past several years, a number of plastic capture technologies have emerged as a means of removing plastic and other debris from the environment or preventing them from finding their way there in the first place. The Great Lakes Plastic Cleanup provides an important opportunity to pilot these technologies on a large scale and under a range of geographic and climatic conditions. While many plastic capture devices were initially developed for use in marine or ocean environments, the initiative highlights their ability to address similar plastic pollution issues in freshwater.

While it is no surprise that the use of technology can be an integral part of the success of any initiative working to address important environmental challenges, the Great Lakes Plastic Cleanup is proving that technology can also act as an effective linchpin for driving broader systems change, including that related to plastic pollution. The initiative has shown that the value of these plastic capture technologies goes well beyond their ability to clean up plastic from the environment when combined with a focus on the collection of locally-relevant data that can further inform regulatory and policy decisions and community outreach and engagement efforts. The success of the Great Lakes Plastic Cleanup is also directly tied to its ability to provide a space for collective action on the part of its strong network of groups and organizations that are working together to end plastic waste.

Given the magnitude of the challenge, it is clear that the plastic pollution issue will not be solved by any one individual or organization alone. Systems change—the identification of organizations and individuals already working on an issue and joining forces to achieve a common goal[12]—has long been identified as a means of moving beyond innovation to lasting, systems-level solutions. It is an effective way to address a complex environmental challenge by drawing upon the strengths and assets of diverse actors in a system. The Great Lakes Plastic Cleanup has proven to be a catalyst for action since its inception, uniting a wide range of like-minded organizations to address an issue of shared concern.

The initiative's network continues to grow and includes partners, collaborators and funders from all orders of government, industry, academia and students, technology providers, not-for-profit and watershed management organizations, local marina and community groups, the artistic community, and the general public (Table 1). While 37 collaborators were involved in its initial year, including marina partners on the Canadian side of the Lakes, the Great Lakes Plastic Cleanup network has since

[12] Walker, J. 2017. Solving the World's Biggest Problems: Better Philanthropy Through Systems Change. Stanford Social Innovation Review. Retrieved from: https://ssir.org/articles/entry/better_philanthropy_through_systems_change#.

Table 1. Great Lakes Plastic Cleanup Network.

Sector	Canada	The United States of America	Binational	Total
Marinas	32	6	0	38
Academia	5	6	1	12
Government	2	2	0	4
Industry	2	3	0	5
Local Community Group/Civil Society Organization	5	5	0	10
National Oceanic and Atmospheric Administration (NOAA) National Sea Grant Network	0	4	0	4
State and Municipal Parks	0	4	0	4
Port Authorities	1	1	0	2
Rotary Clubs	2	1	0	3
Technology Providers	0	0	3	3
Other	3	1	0	4
Total	52	33	4	**89**

expanded to include 89 collaborators in Canada and the U.S., representing a wide range of sectors.

Through the combination of its elements, the initiative provides the flexibility necessary to successfully complement and build on a variety of important works that are underway to address the systemic causes of plastic pollution. This model ensures that important community efforts on the ground, including local beach cleanups and awareness campaigns, happen in step with the types of broader policy and engagement activities necessary to create longer-term solutions.

Piloting technology in the Great Lakes

In its first two years, the Great Lakes Plastic Cleanup focused specifically on piloting two plastic capture technologies, Seabins and LittaTraps, to capture plastic before it enters the lakes or to clean up the waste that has already found its way there. Developed in Australia and New Zealand, respectively, early applications focused primarily on combating the ocean plastic issue. The Seabin acts as a "trash skimmer" and is designed to be installed in the water where it can help remove the plastic and other debris floating on the surface. It can intercept macroplastics and MPs, even including microfibres with an additional filter. Enviropod's LittaTrap is a patented catch basin basket that sits inside a storm drain to prevent litter and other debris from entering the storm drain system. The trap's mesh basket is designed to capture and retain 100% of plastic and other debris over 5 mm. It has high hydraulic conductivity, allows water to pass through it easily, and is both lightweight and structurally robust.

The decision to initially use Seabins and LittaTraps for the Great Lakes Plastic Cleanup was based on several factors including:

- Performance History and Leveraging Past Successes: The initiative looked to build on existing pilot projects that had proven successful results related to the amount of litter that could be collected.
- Cost: The cost of comparable technologies was explored to determine which technologies would provide value for money and contribute to the scalability of their implementation over time.
- Maintenance Requirements: Maintenance requirements for each technology were considered, given the objective of ensuring marinas could easily participate in the project.
- Availability: Technologies were required to be available in Canada to ensure they could be sourced and installed.

While not the first to pilot plastic capture technologies in a freshwater environment, the scale of the Great Lakes Plastic Cleanup has provided an opportunity to determine their viability in unique geographic and climatic conditions that are found across the Great Lakes. The findings point to the role that the appropriate placement of the technologies can play in their effectiveness and the fact that the amount and type of debris captured can vary dramatically by location. The number of plastic capture technology offerings on the market has increased significantly since the launch of the Great Lakes Plastic Cleanup, and the initiative's model allows for the continued incorporation of these in an effort to test their effectiveness in addressing plastic pollution in freshwater and, more specifically, in the Great Lakes.

The current season will see the Great Lakes Plastic Cleanup pilot three additional technologies in the U.S.—the BeBot, Pixie Drone, and Gutter Bin—in an effort to broaden understanding of how different plastic capture technologies operate within a freshwater context. Developed by NITEKO and distributed Poralu Marine, the BeBot is a beach-cleaning robot that mechanically sifts sand and rakes seaweed to remove the plastic and other debris without harming the local environment. It has a sieve capacity of 100 L and can clean up to 3,000 m^2.

The Pixie Drone, developed by Ranmarine and also manufactured by Poralu Marine, is a remote-controlled mobile plastic capture device which is deployed in the water and is equipped with a video camera and LIDAR technology to ensure it avoids obstacles. The Pixie Drone targets any type of floating debris, including organic, plastic, glass, metal, paper, and rubber with a collection capacity of 160 L.

Developed by Frog Creek Partners, the Gutter Bin® stormwater filtration system is a catch basin filter that removes a broad spectrum of pollutants, including plastic, within a drop inlet, a curb inlet, or a trench drain. It has an adjustable frame and funnel that directs polluted water into a Mundus Bag® water filter, customizable according to the specific needs of the location. The Gutter Bin incorporates a backflow preventer to ensure zero pollution loss and an adjustable overflow which allows high-flow events to bypass the filter system.

Enabling data collection and analysis

Studies concerning the prevalence, transport, and fate of plastics in the environment have increased over the past decade, thanks to computer modelling and data collected through beach cleanups and *in situ* samplings. However, those specific to the Great Lakes remain limited, and many models are built from an understanding of how waste is transported in an ocean environment.[13] There is a clear need for a more robust dataset, particularly clarity of data that is related to the amount, location, and types of plastic that are found across the lakes. The use of plastic capture technologies is an effective way to complement these models with additional data.

The advancement of technology has played a major role in opening up new opportunities for deeper interaction and participation in science and research efforts. Technologies and tools for sharing information have allowed community scientists to engage in the collection of data in new and innovative ways and to provide improved means for decision-makers, academia, industry, not-for-profit organizations, and local communities to use the data generated.

When the use of plastic capture technologies is combined with the collection of data following a set of standardized protocols, it plays a critical role in building out a more fulsome picture of the plastic pollution issue. By collecting information about the types and amounts of plastic, along with where they are found, we can begin to build an understanding of potential plastic pathways and sources. This contributes to addressing information and data gaps specific to the Great Lakes which in turn can help to inform policymakers about effective solutions to prevent litter and to aid in the development of locally-relevant mitigation strategies.

The Great Lakes Plastic Cleanup network has been following two protocols developed by the University of Toronto Trash Team for the quantification of debris collected by the plastic capture devices:

- Daily quantification: The purpose of the daily waste characterization is to collect the minimum data necessary to estimate the amount of anthropogenic debris that is being diverted from the Great Lakes.
- Detailed characterization: The purpose of the deeper dive for debris collected over 24 hours is to quantify and characterize what is collected by the plastic capture devices. Data is synthesized and aggregated in an effort to understand the total impact across the network's sites.

Together, these approaches allow for an understanding of how much is reported to be diverted along with the specific types of materials that are captured at each location.

Education and awareness

Numerous studies support the idea that overall motivation and engagement in learning is enhanced by the implementation of technology. More specifically, technology engages individuals behaviourally (more effort and time spent participating in learning activities), emotionally (positively impacting attitudes and interests), and cognitively (mental investment to comprehend content). In addition, providing

[13] Hoffman, M.J. and Hittinger, E. 2017. Inventory and transport of plastic debris in the Laurentian Great Lakes. Marine Pollution Bulletin 115: 273–281. Retrieved from: https://pdfs.semanticscholar.org/80c1/80a30b73fe4fd3490b7fa3355c36c346015a.pdf.

opportunities to embed community engagement within research and monitoring programs has also been shown to ensure their longevity and ultimately, their ability to contribute in a meaningful way to policy development, and the building of more comprehensive datasets. Clearly understanding local priorities and realizing what they are looking to take away from their participation is critical to long-term success. Even a well-designed program may not result in successful engagement if it fails to connect and align with the needs or interests of the local community.

The use of plastic capture technologies as part of the Great Lakes Plastic Cleanup provides an excellent opportunity to bring marina users and local communities together to experience plastic pollution firsthand. The initiative has shown that Seabins, in particular, can be an amazing tool for educating and engaging. The technology offers a way to take immediate action while having the bins visible at the marina provides an opportunity to initiate important conversations about plastic waste in the lakes. One of the most powerful aspects of the use of these technologies is their ability to provide unique opportunities for individuals and communities to generate their questions, contribute to the collection of their data, and advocate for the change they wish to see. It allows those involved to gain a deeper understanding of their natural surroundings and to build an informed public that can advocate more successfully for the protection of human health and their environment. It is also an important way for governments and other social institutions to directly interact with the public while promoting an open collaboration in science, research, and policymaking.

The Great Lakes Plastic Cleanup has also established a unique partnership with the Toronto Zoo to successfully deploy plastic capture technologies on-site, serving as an excellent means of engaging more than 1.2 million visitors to the zoo each year in discussions around the rampant plastic pollution in the Great Lakes.

Key Learnings Detailing the Future of the Great Lakes Plastic Cleanup

Program outcomes and feedback solicited from participating marinas and collaborators point to important lessons related to the piloting of plastic capture technologies in a freshwater environment, the collection of debris for characterization, effective community engagement, and the value of participation in a program of this nature. The findings also highlight a number of areas where further research and action may be necessary and describe implications for effective governance that are related to addressing plastic pollution in the region and beyond.

Piloting plastic capture technologies

The effective placement and installation of the plastic capture technologies was a learning process with equipment that needed to differ dramatically by location. Feedback from marinas has indicated that the specific placement of the technologies corresponds directly to the amount of waste that is retrieved and diverted. Locations with too much wave action during wind events necessitated the suspension of Seabin use for extended periods and in some cases even required its relocation. A number of marinas, including those located in the Georgian Bay region, have experienced challenges with pine needles getting through the Seabin's catch bag and clogging the bottom of the inner bin which contributed to mechanical issues.

Experience with the technology over the first two seasons of the Great Lakes Plastic Cleanup will inform the approach to their continued operation moving forward with improvements being made to equipment location and maintenance practices. The need for a thorough assessment of the site, its topography, and local plant species, prior to the installation of the capture technologies, was identified as critical to finding an installation location where debris collection is maximized while also avoiding unnecessary maintenance issues. Marinas also note that ensuring the technology is installed in a location where it can serve as a focal point increases opportunities for visitors to engage, thus providing additional educational opportunities. Pairing this with visible signage at each location explains how the working of the equipment is also critical for furthering engagement with the local community.

MPs are ubiquitous in the Great Lakes

The amount of debris collected by plastic capture technologies could differ dramatically by day, depending on a wide range of factors including the location of the marina, where the technologies are installed on-site, and weather conditions. In 2021, participating marinas on the Canadian side of the lakes reported diverting close to 15 kg and 67,529 pieces of plastic, or an average of 28 g and 132 pieces per Seabin each day. The devices continue to collect debris even on days when data is not reported which means the actual amount of plastic they remove is substantially greater.

Data collected over the past two seasons is also contributing to building an increasingly detailed plastic profile for each of the lakes, mapping trends on types and sources, and informing the initiative's efforts to prevent plastic pollution and remove that which has found its way into the environment. In 2021, the distribution of plastic found at participating sites on the Canadian side of the lakes was as follows:

- Lake Superior (1 site): 2,492 pieces weighing ~ 547 g
- Lake Huron (11 sites): 2,8591 pieces weighing ~ 6.29 kg
- Lake Erie (2 sites): 1,152 pieces weighing ~ 265 g
- Lake Ontario (6 sites): 41,791 pieces weighing ~ 8.1 kg

MPs make up the majority of the debris collected by the Great Lakes Plastic Cleanup and its network with the top three items found in 2021 being small foam pieces (45.4%), small hard plastic fragments (14.2%) and small pre-production plastic pellets (12.3%). Given the substantial growth of the Great Lakes Plastic Cleanup network this season, including an expansion into the U.S., the overall amount of plastic diverted by the initiative is anticipated to increase significantly while also capturing data from Lake Michigan for the first time.

A key finding from the Great Lakes Plastic Cleanup is the connection between the organic material collected and the amount of small anthropogenic debris. Floating algae and other plant materials that are common in shallow and sheltered waters, like those found at many marinas, collect and accumulate small plastic and other litter. This is an important consideration moving forward as a number of marinas have reported finding primarily organic material in their bins. However, upon closer investigation during the detailed characterization of debris, many small plastic pieces and fragments were found trapped within this vegetation. This highlights the importance of ensuring that the organic material collected in the plastic capture technologies is disposed of appropriately rather than being returned to the lake.

Collaboration with local community groups is critical to successful data collection and engagement

The COVID-19 pandemic had a direct impact on the ability of local groups to get involved in data collection efforts during the initiative's first two seasons. As a result, the debris from the Seabins and LittaTraps required shipment to a number of academic partners for characterization off-site or in their labs. Those locations where local waste characterization partners were in place had more successful data collection efforts as they were able to interact directly with marinas. The importance of local relationships and the opportunity to visit in-person to collect samples played an important role in ensuring that debris was ready for characterization and analysis and also helped with communicating any challenges with operating and maintaining the technologies. This knowledge has directly informed the Great Lakes Plastic Cleanup's efforts to continue building and expanding its waste characterization network. Beginning with an initial three waste characterization partners, including academic institutions and local community groups, the initiative now boasts more than 33 in Canada and the U.S.

Building stewardship

Great Lakes Plastic Cleanup collaborators have noted that the fundamental value of participating in the initiative is based on its ability to provide boaters and residents of their communities with the sense that the marina or municipality truly understands their needs and recognizes that small changes make a great impact over time. Many have indicated that prior to involvement in the initiative, community members often voiced concerns about litter and waste that would accumulate in the water. The plastic capture technologies provide them with a practical means of furthering their commitment to the environment through tangible action.

In the time that we had our Seabins in the water last summer, we were able to help create a heightened environmental awareness in individuals of all ages! Our customers would stop dead in their tracks to ask questions about the Seabin and how it was helping clean the water. They loved walking by to see its contents and watching it remove any oil sheen from the water's surface.

Great Lakes Plastic Cleanup marina, personal communication, January, 2021

Marinas located downstream of urban centres or areas of increased activity along the waterfront, in particular, appreciated the use of novel plastic capture technologies to address waste at their site and also pointed to litter being a major concern in these locations.

In addition to marina partners, the Great Lakes Plastic Cleanup has also provided an important opportunity to build on the important efforts to address plastic pollution underway by a wide range of collaborators. Moving forward, strategic partnerships will further strengthen the findings and outcomes of the initiative and with this in mind, opportunities will be explored to continue partnering with new sponsors, local community groups, and audiences. The Great Lakes Plastic Cleanup provides a range of opportunities for tailored employee engagement and on-the-ground participation by governments, industry, academia and other organizations.

We see value in being a part of the Great Lakes Plastic Cleanup because we are all working towards the same goal—removing plastics that can, or already are, contaminating our water. The more communities that get on board with curbing their own impacts on the quality of our water, the faster we can actually make and see a difference. Great Lakes Plastic Cleanup collaborator, personal communication, January, 2021.

The initiative also inspired a sister project in B.C.—the Vancouver Plastic Cleanup—led by Swim Drink Fish which was built on the model developed by the Great Lakes Plastic Cleanup. This highlights the role that the initiative and its model can play in addressing plastic waste across the country based on its flexibility in adopting local considerations, interests and the needs of all collaborators and partners while ensuring it is scalable and practical.

Policy development

The Great Lakes Plastic Cleanup has already contributed directly to the development of a much-needed policy to address plastic pollution, including through Ontario's Bill 228, Keeping Polystyrene out of Ontario's Lakes and Rivers Act, 2021. There has also been a draft regulation to address stray pre-production pellets at industrial facilities in collaboration with the Chemistry Industry Association of Canada—which is responsible for the implementation of Operation Clean Sweep (OCS) in Canada—and renowned academics. Members of the Great Lakes Plastic Cleanup also contributed to the development of OCS.

On 5 November 2020, MPP for Parry Sound—Muskoka, Norm Miller, introduced a Private Member's initiative, Bill 228. The legislation aims to help limit MP pollution by requiring that all new dock floats and buoys made from polystyrene are fully encapsulated to prevent the foam from breaking down and entering aquatic ecosystems. The legislation was debated in the Legislature on 23 February 2021, and the efforts of the Great Lakes Plastic Cleanup in contributing to important research on the issue through the use of plastic capture technologies was singled out.

Efforts underway as part of the Great Lakes Plastic Cleanup are also helping to inform the development of a proposed regulation to prevent pre-production plastic pellets from entering the environment. Despite mounting evidence of the presence and impacts associated with MPs in freshwater environments, there are currently no mandatory requirements for the control of their release from industrial facilities. A regulatory requirement aimed at complementing efforts under the OCS program can reduce plastic pollution and the loss of valuable plastic materials.

The creation of a regulation under the Ontario Environmental Protection Act requiring the plastics industry to install zero-loss containment systems and report on the leakage of pellets from their facilities has the potential to encourage further measurable action to prevent their release into the environment, while also increasing accountability. The proposed regulation would outline a solution that builds on the OCS program by incorporating more stringent requirements, acknowledging industry leadership demonstrated by members of the CIAC Plastics Division, and providing them with an exemption from its full requirements. In this way, the regulation could support the broader uptake of OCS across the province, nationally, and globally and ensure that resin and plastic manufacturers beyond CIAC members are also required to be a part of the solution.

Conclusion

Experience from the Great Lakes Plastic Cleanup since its launch in 2020 illustrates the need for much more work to end plastic pollution. For this reason, the initiative has continued to expand with marinas and collaborators from across the region contributing to the prevention and removal of plastic waste entering the lakes. Each new plastic capture device installed or collaborator joining the network increases the amount of plastic removed from the Great Lakes and its waterways. However, the true value of the initiative extends well beyond capturing plastic.

The Great Lakes Plastic Cleanup has effectively highlighted that technology and tools for sharing information can play a critical role in driving the systems change which is necessary to eliminate plastic pollution. While the implementation of its model has contributed to important lessons specific to plastic, the unique combination of the initiative's elements also has the potential for broader application and tangible results across a wide range of significant environmental challenges.

A familiar refrain when discussing environmental issues is that you cannot manage what you cannot measure. At the same time, environmental challenges are often prioritized based on where information and data already exist. As such, the Great Lakes Plastic Cleanup's efforts to increase the amount of available data related to sources, types, and quantities of plastic in the region are playing a key role in ensuring that the importance of addressing plastic pollution in the region, and freshwater systems, in general, remains a top priority.

In addition, research has shown that when people are provided with the opportunity to solve a challenge with their own hands, the outcomes are of much greater value to them. For example, collecting litter and discussing it significantly increases the likelihood that it will be disposed of appropriately in the future. The Great Lakes Plastic Cleanup's focus on engaging local communities firsthand in activities that give them direct exposure to what happens to plastic when it enters the environment can inspire the adoption of a new behaviourally-based approach to addressing plastic pollution.

Moving forward, the Great Lakes Plastic Cleanup will focus on continuing to mobilize and translate the knowledge gained through the initiative while thoughtfully growing and expanding its efforts to reach new locations and audiences and deepen its contributions to ending plastic pollution in the lakes. This will involve pursuing a number of interrelated strategies:

- Expanding to New Locations: In addition to continuing to strengthen collaboration and support for its current sites, the initiative will look to expand to strategic locations across the Great Lakes. This may include areas considered to be plastic "hot spots", locations that are ecologically sensitive or where new audiences could be engaged (e.g., indigenous communities or local education organizations).
- Exploring New Opportunities for Collaboration: The Great Lakes Plastic Cleanup benefits from collaboration across its dedicated network. Strategic collaborations will further strengthen the outcomes of the initiative and with this in mind, opportunities will be explored to partner with new sponsors, local community groups, and other organizations committed to addressing plastic pollution.

Informing Broader Solutions: While addressing the plastic that has already found its way into the lakes is of critical importance, preventing it from entering the environment in the first place is an important goal of the Great Lakes Plastic Cleanup. The initiative has already contributed directly to the development of much-needed policy and has continued to pursue opportunities to inform solutions for addressing plastic waste along the value chain in collaboration with government and industry partners.

It is clear that it will take a suite of coordinated actions over the long term to effectively address the systemic causes of plastic pollution and its associated impacts on our critical freshwater ecosystems. Through its unique combination of elements, the Great Lakes Plastic Cleanup is showing that practical action can be taken now that will have a measurable and immediate impact. The learnings and outcomes outlined in this chapter can be used to scale and replicate its successes and facilitate further action across watersheds while informing effective decision-making to address the significant challenge of plastic pollution.

References

Brookson, C.B. et al. 2019. Microplastics in the diet of nestling double-crested cormorants (Phalacrocorax auratus), an obligate piscivore in a freshwater ecosystem. *Canadian Journal of Fisheries and Aquatic Sciences* 76: 11.

Canadian Council of Ministries of the Environment. 2020. Canada-Wide Action Plan on Zero Plastic Waste: Phase 2. Retrieved from https://ccme.ca/en/res/ccmephase2actionplan_en-external-secured.pdf

Chae. Y. et al. 2018. Trophic transfer and individual impact of nano-sized polystyrene in a four-species freshwater food chain. *Scientific Reports* 8: 284.

Government of Canada. 2018. Ocean Plastics Charter. Retrieved from https://www.canada.ca/en/environment-climate-change/services/managing-reducing-waste/international-commitments/ocean-plastics-charter.html

Governments of Canada & U.S. 2019. 2019 Progress Report of the Parties. Retrieved from https://binational.net/wp-content/uploads/2019/06/Final-2019-PROP-English-June-7.pdf

Hoffman, M.J. and Hittinger, E. 2017. Inventory and transport of plastic debris in the Laurentian Great Lakes. *Marine Pollution Bulletin,* 115: 273–281. Retrieved from: https://pdfs.semanticscholar.org/80c1/80a30b73fe4fd3490b7fa3355c36c346015a.pdf

McNeish, R.E. et al. 2018. Microplastic in riverine fish is connected to species traits. Scientific Reports, 8, 11639.

Peller, J. et al. 2021. Sequestration of microfibers and other microplastics by green algae, Cladophora, in the US Great Lakes. Environmental Pollution, 276. 116695.

Pollution Probe and the Council of the Great Lakes Region. 2018. Great Lakes Plastics Forum: Solutions for a Sustainable Future Summary Report. Retrieved from https://www.pollutionprobe.org/wp-content/uploads/Great-Lakes-Plastics-Forum-Summary-Report.pdf

Province of Ontario 2017. Ontario's Great Lakes Strategy. Retrieved from https://www.ontario.ca/page/ontarios-great-lakes-strategy#section-3

Rochman, C.M., Hoh, E., Kurobe, T. and Teh, S.J. 2013. Ingested plastic transfers hazardous chemicals to fish and induces hepatic stress. *Scientific Reports* 3: 3263.

Walker, J. 2017. Solving the World's Biggest Problems: Better Philanthropy Through Systems Change. Stanford Social Innovation Review. Retrieved from: https://ssir.org/articles/entry/better_philanthropy_through_systems_change#

Xia. X. et al. 2020. Polyvinyl chloride microplastics induce growth inhibition and oxidative stress in Cyprinus carpio var. larvae. *Science of the Total Environment* 716: 136479.

Chapter 7

Microplastic Sources, Contamination, and Impacts on Aquaculture Organisms

H.U.E. Imasha and *S. Babel**

INTRODUCTION

Global plastic production has been rapidly growing with higher demand and wider applications of plastic products. In 2018, nearly 359 million tons of plastics have been produced globally (Plastics Europe, 2019). A large fraction of plastic garbage, nearly 6,300 million tons, has been created globally between 1950 to 2015 and has ultimately ended up in the aquatic environment due to improper management and misuse of plastics (Geyer et al., 2017). When released into the environment, large plastics could fragment into microplastics (MPs) (Peng et al., 2020). Particles that are less than 5 mm in diameter are generally defined as MPs. These can also be manufactured as pellets in industry or scrubbers in cosmetics and are a very important component in plastic pollution (Pinheiro et al., 2017). MPs are widespread, and they have been detected in beaches, sea bed sediments, surface water, freshwater, wastewater effluents, indoor and outdoor environments, and even in Arctic and Antarctic seas (Karbalaei et al., 2018).

Owing to their tiny sizes, MPs become accessible for ingestion to a large range of organisms and shift to higher trophic level organisms across food chains (Chagnon et al., 2018). Many studies have demonstrated ingestion of MPs by a wide variety of marine organisms, including plankton (cope pods, euphausiacea (krill) and larval stages of mollusks, and decapods) and other invertebrates (polychaetes, bivalves, echinoderms, and decapods) (Van Cauwenberghe and Janssen, 2014). The ingestion of MPs does not just result in physical harm to the animals but also can cause toxicity due to chemicals, such as additives leaching out from MPs can end up in organism

School of Bio-Chemical Engineering & Technology, Sirindhorn International Institute of Technology, Thammasat University-Rangsit Center, 99 Moo 18, KhlongLuang, PathumThani 12120, Thailand.
* Corresponding author: sandhya@siit.tu.ac.th

tissues (Hahladakis et al., 2018). It has been found that ultimately MPs can cause damage to organisms and ecosystems and even affect human health (Karbalaei et al., 2018).

According to Food and Agriculture Organization (FAO), there is unprecedented growth in world aquaculture production with a high record of 114.5 million tons (live mass) in 2018. Increased production has led to increased consumption of aquaculture products globally. Aquaculture products provide valuable sources of nutrients for a healthy diet. The aquaculture sector is an important source of animal protein intake by the human population. Hence, it plays an important role in the growth of the human population (FAO, 2018).

Researchers have shown that MPs appear in aquaculture sites as they are generally situated in areas where MPs cannot be easily drained out to open seas,thus resulting in higher bio accumulation in aquaculture organisms than in wild organisms. Mathalon and Hill (2014) reported a significant accumulation of MPs in farmed mussels (178/items per individual) compared to wild mussels (116/items per individual). The ingestion of MPs can cause physical damage to the organisms while these MPs in contaminated aquaculture products may pose a health risk to humans when consuming such products. Therefore, it is vital to investigate profusion and characteristics of MPs presented in aquaculture species to protect human health.

This chapter summarizes sources of MPs in the aquaculture environment, the mechanisms of interactions with organisms, contamination,and their effects on biota. It also reviews the published literature regarding the occurrence of MPs in farmed, wild environments, and markets mussels,given that mussels have been used as the sentinel organism to track pollution created via MPs.

Aquaculture Production

World aquaculture production

Food and Agriculture Organization (FAO) defines aquaculture as "cultivation of aquatic animals, such as finfish, crustaceans, molluscs, etc., and aquatic plants mostly algae, utilizing or within freshwater, seawater, inland saline water and brackish water" (FAO, 2020). In 2018, the global aquaculture sector reached a record high of 114.5 million tons of aquaculture production in a live mass which had a commercial value of USD 263.6 billion. In 2018, the contribution to the world farm food fish production only from inland aquaculture was 62.5% (41.3 million tons of aquatic animals). The cultivation of aquatic animals in 2018 was dominated by the finfish produced in coastal aquaculture, mariculture, and inland aquaculture. In addition to finfish, mollusks, bivalves, marine invertebrates, aquatic turtles, and frogs have been grown as well. Seaweed is the dominant farmed aquatic algae in the world of aquaculture production. Asia accounts for 89% of global aquaculture production of cultivated aquatic animals. China, India, Egypt, Bangladesh, Indonesia, Norway, Vietnam, and Chile are among the top producers. Additionally, the fisheries and aquaculture sector represented employment opportunities for 59.5 million people in 2018 which is an incremental increase compared to 2016 (FAO, 2020).

Use of plastic in fisheries and aquaculture

There is an increasing demand for aquaculture products (especially fish) with the increasing growth of the human population. The aquaculture sector plays a significant role in meeting the demand for fish, shellfish, and other aquaculture products as a result of the growing consumer market (Lusher et al., 2017). The use of a massive quantity of plastics for packaging and fabrication of various equipment has become a very common practice in the aquaculture sector. Characteristics like its resistance to abrasion, durability, rust resistance, light weight, low cost, ability to mold into specific shapes, and improvement of longevity and reliability of equipment have made plastics an excellent material for use in an aquatic environment. Various types of plastics are utilized in mariculture to inland aquaculture systems to fabricate equipment, such as cages, ponds, tanks, and ropes (Huntington, 2019). Low Density Polyethylene (LDPE), High Density Polyethylene (HDPE), Polyvinyl Chloride (PVC), Polypropylene (PP), Polystyrene (PS), Polyamide (Nylon), Polycarbonate (PC), Acrylic (PMMA), and Fiber Reinforced Plastics (FRP) are some of the most common plastic types that are used in aquaculture systems (Biernacki, 2011).

Sources of MPs in Aquaculture Systems

While plastics are a major component in the aquaculture sector, a considerable portion of equipment and other materials used in aquaculture end up in marine and other aquaculture environments as waste. Sources of MPs in aquaculture systems can be classified into direct and indirect sources.

Direct sources

In aquaculture systems, when materials composed of plastics, such as commercial fishing gear, net cages, plastic lines, and trammel nets, are continuously used for a longer period, they are exposed to direct UV rays, tidal action, abrasion, and temperature variations that contribute to its embrittlement and fragmentation. These materials are no longer suitable for use after their breakdown. Presently utilized artificial polymers do not immediately decompose and take a long duration to decompose and finally produce MPs. A large portion may end up becoming marine or freshwater debris. Browne et al. (2015) and Nelms et al. (2017) reported the presence of fisheries and aquaculture wastes in marine garbage. Cózar et al. (2014) reported the presence of fisheries and aquaculture wastes floating on surface water, while Iñiguez et al. (2016) reported its presence on the sea floor. The Republic of Korea has estimated the national aquaculture debris input to the marine environment as 44,081 tons from lost fishing gears, 2,374 tons of garbage input from fishing boats, and 4,382 tons from expanded polystyrene (EPS) floats that have been lost from aquaculture facilities (Lee et al., 2014). Castro et al. (2016) believe that the high levels of synthetic fibers in Jurujuba Cove, Brazil, originated from nearby local mussel farming. Davidson (2012) has observed the release of MPs in Polystyrene (PS) floats has been utilized in aquaculture facilities and docks by boring isopods (*Sphaeroma* spp.).

Another main and direct source of MPs to the aquaculture system,especially mariculture systems, is abandoned, lost, or otherwise discarded fishing gear (ALDFG). Severe weather conditions, accidents, or conflicts with other users and wearing and tearing of anchor ropes may deteriorate aquaculture structures. These ALDFG has also been identified to cause marine pollution (Lee et al., 2015). Shellfish culture production is a key contributor to coastline debris, including EPS floats, plastic net sheets, bags, ropes, and baskets (Liu et al., 2015; Jang et al., 2014a,b; Lee et al., 2015). Bendell (2015) found the loss of anti predator nets from the shellfish industry to be a source of plastic contamination in aquaculture systems. Floerl et al. (2016) referred to the release of net and rope fibers in the form of MPs into the environment as a result of the continual removal of biofouling organisms from aquaculture facilities. Bendell (2015) reported the pollution of Puget Sound by shellfish aquaculture, and Andréfouët et al. (2014) detected ALDFG from the pearl oyster aquaculture in French Polynesia. These are two examples of plastic waste being released by the aquaculture industry.

Indirect sources

Wind and atmospheric fallout results in the dissemination of MPs in aquaculture environments. In the Gulf region, due to its semi-closed geographic situation, typhoons increase inputs of MPs in the Gulf and consequently lead to an increased concentration of MPs in farmed oysters (Wang et al., 2019). Moore et al. (2002), Lattin et al. (2004), and Law et al. (2014) pointed out how storms and rainfall contributed to increases in concentrations of MPs in marine. Fischer et al. (2016) and Kukulka et al. (2012) identified that wind could change the distribution of MPs in freshwater lakes. Rivers and runoff are other indirect sources that bring MPs into aquaculture environments. MPs in facial care products are released into receiving waters and are accumulated in aquaculture systems. The concentration of MPs is significantly higher at the entrance of a fishpond than at the outlet, thus showing the accumulation inside the aquaculture system and the presence of MPs in the sediments (Bordós et al., 2019). MPs can enter rivers through the direct discharge of treated sewage from wastewater treatment plants. Although treatment plants can remove MPs from treated sewage around 98%, the remaining MPs particles are still directly discharged into the receiving water (Murphy et al., 2016). Runoff can also add MPs enclosed in soil into the river. The commercial seafood production industry also contributes to the accumulation and dispersal of MPs in aquatic environments through the direct disposal of plastic materials. The number of discarded tools used worldwide between 2000 and 2010 was approximately 10 million tons per year that was potentially accumulated by benthic species (Pauly and Zeller, 2016). Each year fish meal and fish oil is produced from the small pelagic species which mistakenly consume MPs as their food (To et al., 2016). A proportion of this fish meal is used as feed in aquaculture industry. These pre-contaminated feeds with MPs can be another source of its entrance in the aquaculture industry, although there is no evidence of this at present. Contaminated bait may result in either introduction of MPs to the focus organisms upon catch or to the environment as bait is discarded or lost (Lusher et al., 2017).

Distribution of MPs in the aquaculture environment

Today, we have come to a "Plastic Age" that has changed our life massively due to the advantages and convenience of using plastic materials. Production of plastics is increasing worldwide annually (Plastics Europe, 2019). In contrast, plastic pollution was observed and reported in all the ecosystems in the world (Gall and Thompson, 2015). Distribution of MPs was reported in freshwater (Ta et al., 2020), oceans (Peng et al., 2020), treated wastewater (Dris et al., 2015), lakes sediments (Fischer et al., 2016), soil organisms (Huerta Lwanga et al., 2016), air (Gasperi et al., 2015), and marine and freshwater animals (Karami et al., 2017; Phillips and Bonner, 2015). Studies presenting data on MP pollution in aquaculture sites are scarce at this moment. Table 1 summarizes the distribution of MPs in aquaculture sites. All the compartments, including water, sediment, and biota, and in most of all aquaculture environments around the world are contaminated with MPs.

Interactions of MPs with Aquatic Organisms Under Laboratory Conditions and Field Environment

MPs have been discovered widely around the globe. Different trophic level organisms interact with MPs through a variety of routes by which exposure and interaction may take place. Lower trophic level organisms which were exposed to MPs present in the environment (water or sediment) and were mistakenly ingesting them, became an indirect source of MPs and later contaminated prey for their predators through trophic transfer. This phenomenon raises concerns about the interactions of MPs with biota, thus inspiring several exposures and toxicological studies in laboratory conditions. Uptake, movement, and dissemination of artificial particles in entire organisms and removed tissues (e.g., gills, intestinal tract, and liver) can be easily monitored under laboratory conditions. Laboratory experiments have proven that a variety of marine biota, across the different trophic levels, can consume or absorb MPs easily (Lusher et al., 2017). Protists, cope pods, annelids, echinoderms cnidaria, amphipods, molluscs, fish, and birds have been reported to be contaminated with MPs (Christaki et al., 1991; Cole et al., 2015; Wright et al., 2013; Nobre et al., 2015; Hall et al., 2015; Ugolini et al., 2013; Avio et al., 2015; Pedà et al., 2016; Tanaka et al., 2013). Several studies have reported interactions of freshwater biota with MPs. In comparison, very few studies have reported the ingestion of MPs by free-living terrestrial organisms; however, in a laboratory study, Rodriguez-Seijo et al. (2017) have indicated the ability of earthworms to consume plastic particles present in the soil.

Mechanisms of MPs-biota interactions under laboratory conditions

Uptake, accumulation, and toxicity of contaminants can be effectively studied using laboratory methods (Qu et al., 2018). Mechanisms of biota-MPs interactions have been explored in many studies under laboratory conditions. Mainly, there are two main ways of MPs-biota interactions: adhere and uptake. Cole et al. (2013, 2015) report that micro- and nano-plastics are able to adhere to the external limbs of cope

Table 1. Distribution of MPs in Aquaculture Environments.

Aquaculture Environment	Site/Aquaculture Activity	Compartment Studied	Identification	Size	Reference
Estuary	Qin River, China	Surface water and Sediment Water	Fibers, Lines, Sheets, Fragments, and Foams	1 to 5 mm	(Zhang et al., 2020)
Channel	Baynes Sound, Shellfish Growing Region of Canada	Surface Sediments	Fibers	Less than 0.65 μm	(Covernton and Cox, 2019)
Intertidal Zone	Shellfish Aquaculture Site in Canada	Shellfish (Clams and Oysters) Surface water and Sediment	Fibers, Fragments, Spherules	Sediments –100 to 499 μm Surface water – 1,000–5,000 μm Clams and oysters – 100 to 499 μm	(Covernton et al., 2019) we compared microplastic particle (MP)
Costal	French Atlantic Coasts. Aquaculture of Mussels and Oysters	Blue Mussels and Pacific Oysters	Mussels – Polypropylene (PP), Polyethylene (PE), Polystyrene (PS), Polyester (PES), Fragments, Filaments, Fibers Oysters – PE, PP, Polyester, andPolystyrene, Fragments	Mussels – 50–400 μm Oysters – 20–100 μm	(Phuong et al., 2018)
Open Sea Water	A Mussel Farm in Germany	Bivalves	0.36 ± 0.07 particles g/1 (wet weight)	5 to 10 mm (50.0%)	(Van Cauwenberghe and Janssen, 2014)
Aquaculture Water and Aquaculture Pond nfluents	Pearl River Estuary of Guangzhou, China	Water Samples	PP and PE	< 1,000 μm (56.3–87.7%) < 333 μm (43.7%)	(Ma et al., 2020)
Bay	Baynes Sound, British Columbia. Shellfish farming industry	Clams	Fibers (90%) Films (5.3%) Fragments (4.7%)		(Davidson and Dudas, 2016)

pods, such as antennules, setae, and swimming legs. Watts et al. (2014) showed that MPs could adhere to the gills of green crabs, and Paul-Pont et al. (2016) reported MPs could adhere to blue mussels. The uptake of MPs through ingestion is the most studied interaction of MPs with biota. Due to the small sizes of MPs, biota cannot distinguish MPs from prey items or ingestion happens accidentally while filter feeding (Bergmann et al., 2015). Direct consumption of MPs is very common among suspension feeders,such as oysters (Bergmann et al., 2015) and mussels (Avio et al., 2015). Direct consumption of MPs is also very common among deposit feeders, e.g., cucumbers (Graham and Thompson, 2009), annelids (Browne et al., 2015) Norway lobster (Welden and Cowie, 2016), and crabs (Watts et al., 2014). Green crabs can mistakenly allow the uptake of MPs during their ventilation through the gills (Watts et al., 2014). These uptake behaviors are mainly due to the indistinguishable nature between food and MPs. Savoca et al. (2016) reported that MPs in the marine environment create a dimethyl sulfide signature which acts as a keystone odorant in pelagic food webs. The diversity of organisms from zooplankton to baleen whales can mistake plastics for prey.

Laboratory experiments have studied biota MPs interactions as well as pathways of MPs within an organism. Watts et al. (2014) and Farrell and Nelson (2013) reported the accumulation of MPs in the digestive tract of green crabs when MPs contaminated blue mussels that were used as food for green crabs feeding under laboratory conditions. Besseling et al. (2014) and Cedervall et al. (2012) showed the trophic transfer of MPs from green algae to the Daphnia magna and then to several species of fish, including Crucian carp, Tench, Bleak, Northern pike, Atlantic salmon, and Rudd. Laboratory experiments have reported the transfer of fluorescent PS from zooplankton to the mysid shrimp (*Mysis* spp.). However, laboratory exposure experiments results should be treated with care as often concentration exceeds more than that observed in the field.

Field Observations of Biota Interactions with MPs

Organisms from a wide variety of environments, including the ocean surface, water column, benthos, estuaries, aquaculture, and beaches, have been observed to have an increased uptake of MPs (GESAMP, 2016). More than 220 species have been observed to ingest MPs debris (UNEP, 2016; GESAMP, 2016). Out of this, 58% of the studied species were commercially targeted species. These organisms include seabirds, marine mammals, shellfish, fishes, etc. (Lusher et al., 2017). Table 2 summarizes the field observation of interactions of commercially important organisms with MPs. Bivalves are the most studied species.

MP Pollution in mussels

Field experiments on MPs in mussels

Mussels have been widely utilized as sentinel organisms around the globe for MP pollution. This is mainly due to the extensive filter-feeding activity of mussels that exposes them directly to MPs in the environment; also, their wide distribution around the globe, vital ecological niches, close relationship with marine predators, and risk

Table 2. Field Observation of Interactions of Commercially Important Organisms with MPs.

Species	**Location**	**Average Number of Particles (*n*) Per *g* Soft Tissue**	**Type of Particles**	**Reference**
MPs in Invertebrates; Crustaceans and Echinoderms				
Common Shrimp (Crangoncrangon)	Belgium, France, Netherlands,and the United Kingdom	0.75 (0.03 to 1.92) n =165 Size: 200–1,000 μm (fibers)	Fibers, Granule, and Film	(Devriese et al., 2015)
Norway Lobster (Nephrops Norvegicus)	North and West Scotland	0.40 mg Maximum – 0.80 mg n = 1,450	Fibers Polyethylene, film, and PVC	(Welden and Cowie, 2016)
Sea Cucumbers	EquitorialMid-Atlantic and Indian Ocean	No Data	Fibers	(Taylor et al., 2016)
MPs in Commercial Fish				
Atlantic Cod (GadusMorhua)	Norway		Nine Different Polymers; PE, PP, PVC, PET (Polyester), and Nylon	(Bråte et al., 2016) mesoplastics (5–20 mm)
Atlantic mackerel (Scomberscombrus)	North and Baltic Sea	n = 138	PE and PA	(Rummel et al., 2016) we investigated 290 gastrointestinal tracts of demersal and pelagic fish species
Atlantic Cod (GadusMorhua)		n = 205	Film/Sheet, Threads, and Fragments	(Liboiron et al., 2016)

to the health of human beings, as mussels are consumed as a whole animal without discarding the digestive organs, make them all more hazardous. Other biotas, which have been reported to ingest MPs, such as fish and crustacean species usually contain MPs in their digestive tracts. Cleaning before consumption will remove most of the digestive tract, the head, and the gills, which often contain a major portion of the MPs (Devriese et al., 2015). Furthermore, the increased uptake of MPs by wild and farmed mussels has been reported widely showing mussels to be highly susceptible to ingesting MPs. So far, field investigations have been mainly done with blue mussels (Mytilus spp.) due to their widespread nature. But some studies focused on *Pernaviridis* and *P. perna* for MP pollution. Numerous countries have carried out field investigations of MPs in mussels around the globe, including the United Kingdom, Belgium, France, Germany, Italy, Turkey, Greece, Netherlands, Portugal, Denmark, Norway, Finland, Canada, China, Brazil, and Indonesia.

Characteristics of MP Pollution in Mussels. Literature studies show that MPs are often detected in both wild and cultured mussels in numerous countries around the globe. Fibers are the most dominantly identified MP morphotypes that are followed by fragments in field studies. Murphy (2018) reported the presence of pellets in mussels. The most reported polymer types are PE, PP, PS, PES, PET, PA, PVC, and cellophane. Some studies demonstrated that the morphology and polymer composition of MPs in mussels appears to be consistent with their adjacent environmental media (Leslie et al., 2017; Li et al., 2018; Qu et al., 2018). Qu et al. (2018) identified a correlation between MPs in mussels and the surrounding seawater reflecting that morphology and polymer types of MPs in mussels can indicate the real contamination in the adjacent environment. Qu et al. (2018) and Digka et al. (2018) showed that mussels can have an uptake of a larger proportion of smaller sizes of MPs, compared to the adjacent environmental media. Digka et al. (2018) reported smaller MPs 62.3% in seawater, 96.9% in sediments, and 100% in mussels in the Northern Ionian Sea. The abundance of MPs in studies varies, ranging between 0.05 items/g to 259 items/g (Bonello et al., 2018; Murphy, 2018). The abundance of MPs in mussels varies due to the degree of background contamination, regional differences in MPs content, and the diversity of methods applied by different researchers. Browne et al. (2010, 2011) suggested a positive correlation between human population density and coastal concentrations of MPs. Furthermore, Li et al. (2016) reported that mussels in areas where intensive human activities happened contained a higher number of MPs which suggested the concentration of MPs in mussels was closely associated with human activities. Karlsson et al. (2017) reported significantly higher concentrations of MPs in mussels (3.7×104 items/kg dry weight) compared to surrounding sediment and seawater, which was 48 items/kg dry weight and 27 items/L, respectively, thus it further indicated accumulation of MPs in mussels. Qu et al. (2018) also revealed a positive and quantitative relationship between MPs in mussels and their adjacent waters.

Farmed Mussels Vs. Wild Mussels. Several studies have analyzed the relationship between farmed mussels and wild mussels. Mathalon and Hill (2014)reported that as the farmed mussels grew on polypropylene plastic lines, they contained more MPs than wild mussels. However, Li et al. (2016) reported reverse results, and they suggested that it was not a single factor such as plastic lines on farms that decided the abundance of MPs in mussels but the level of total pollution by MPs in the environment. However, Vandermeersch et al. (2015) found no significant variation in levels of MPs in farmed and wild mussels.

MPs in Mussels Purchased From Markets. There is a variation in MPs concentration among wild and farmed/commercially-sourced mussels as reported in the literature. Cho et al. (2019) comment that market-based assessment is an appropriate method for the evaluation of human ingestion of MPs through seafood consumption due to two reasons. First is that the majority of people consume seafood by directly buying from the market, and second is that MPs concentrations can alter during transportation from the aquaculture site to the market. Table 3 summarizes the widespread presence of MPs in mussels that are sold in markets in various countries. Italian markets have shown the highest concentration. There is a huge variation in the concentration of MPs in the market-sold mussels due to variations in regional contamination status by MPs and differences in the analytical methods. Bivalves from markets have been reported with a lower concentration of MPs compared to bivalves from coastal regions and aquaculture farm areas; it is mainly due to the depuration of MPs during transportation and storage (Cho et al., 2019). A decrease in the concentration of MPs by 30% in mussels and oysters after 3 days of depuration has been observed by Van Cauwenberghe and Janssen (2014). There is a difference between the abundance of MPs in supermarket-bought mussels based on whether they were pre-processed (either being pre-frozen and chilled or cooked-frozen-chilled) or alive at the time of purchase. Processed mussels have significantly higher MPs concentration compared to live mussels from farmed sources that are sold at supermarkets (Li et al., 2018).

Laboratory experiments on MPs in mussels

Laboratory experiments have been conducted to research the uptake, accumulation, and impacts of MPs in mussels. Mussels have an uptake of MPs mainly through ingestion and adherence to soft tissues, such as the mantle, foot, adductor, visceral tissue, and gonad. Laboratory studies show selectivity characteristics of mussels for a specific size range of MPs during the uptake and egestion process, which is in line with the findings of field studies. Results oflaboratory studies also indicate that there is a positive and quantitative relationship between MPs in mussels and their adjacent waters (Li et al., 2019). Qu et al. (2018) have observed that smaller sizes of MPs are more ingested by mussels that are similar to the field investigations. But the researchers have also reported that mussels can ingest beads easily compared to fibers due to the beads having smaller sizes than fibers,while field studies reported more ingestion of fibers by the mussels. This shows the difference between laboratory exposure studies and field studies. Factors like using short exposure time and neglecting the use of MPs that have environmentally relevant properties in laboratory studies can be the reason for such differences between field and laboratory experiments (Qu et al., 2018).

Table 3. Observation of the Presence of MPs in Mussels from SupermarketsAround the World.

Country	Species	Analytical Method			Concentration of MPs		Common Size (μm)	Dominant Shape	Reference
		Digestion/ Density Separation	Filter Pore Size (μm)	Identification	*n*/*g*Wet Weight	*n*/Individual			
China	*M. galloprovincialis*	H_2O_2/NaCl	5	μFT-IR	2.4[a]	NA	5–250	Fiber	(Li et al., 2015)
France	*M. edulis*	HNO_3:HCl	5	Hot needle	0.06 ± 0.13	NA	NA	Fiber	(Van Cauwenberghe and Janssen, 2014)
Belgium	*M. edulis*	HNO_3:HCl	25–60	Hot needle	0.35[ab]	NA	1,000–1,500	Fiber	(De Witte et al., 2014)
Spain	*M. galloprovincialis*	HNO_3:HCl	5	Hot needle	0.04 ± 0.09	NA	NA	Fiber	(Vandermeersch Van Cauwenberghe et al., 2015)
Italy	*M. galloprovincialis*	H_2O_2	0.45	Microscope	8.33 ± 3.58	3.6–12.4	1,700–1,900	Fiber	(Renzi et al., 2018)
The UK	*M. edulis*	H_2O_2/NaCl	5	μFT-IR	0.9[a]	NA	5–250	Fiber	(Li et al., 2018)
Canada	*M. edulis*	H_2O_2/NaCl	0.8	Microscope	7.42	178	NA	Fiber	(Mathalon and Hill, 2014)
Korea	*M. edulis*	KOH/ lithium meta-tungstate	20	μFT-IR	0.07 ± 0.06	0.77 ± 0.74	100–200	Fragment	(Cho et al., 2019)

a = mean value, b = only fibers have analyzed

Effects of MPs on Aquaculture Organisms

Despite the growing literature on the presence of MPs in the biota, there is little information on its effects on the biota, especially when the research related to the human health risk of MPs is still at the initial stage.

Effects of MPs on biota

After ingestion, physical damages can follow, such as harm to the digestive organs, a false illusion of satiety, and lower nutrient ingestion that can lead to death (Zhang et al., 2019). These ingested MPs can also cause various cellular and molecular level toxic effects. Smaller particles of MPs accumulate inside the organisms and create the path for MPs to enter the food chain and reach higher trophic levels, thus creating further chronic effects (Farrell and Nelson, 2013). In addition, smaller-size MPs and nano-plastics pose severe effects, such as penetration of fungi cell walls and creation of toxicity, that can penetrate the highly selective fish membrane and create more adverse effects (Zhang et al., 2019).

Ingestion of PE fragments by mussels was found to result in histological changes in gills and digestive tissues and hemocytic aggregates in the digestive gland. Ingestion of nano-PS by mussels has resulted in lower filtering activities. Furthermore, ingestion of MPs by mussels results in many adverse effects, including genotoxicity, inflammatory response, alterations of antioxidant system, neurotoxic effects, dysplasia, transcriptional responses, and lysosomal membrane destabilization (Li et al., 2019). Sussarellu et al. (2016) showed an interruption to the reproductive processes and a decreased larval development after oysters were exposed to PS pellets for two months.

Entanglement of MPs into the aquatic biota (crustaceans, seabirds, mammals, and sea turtles) also creates negative impacts, such as drowning, suffocating, and starving, that can lead to death. More than 200 species have been reported to suffer due to entanglement and ingestion of plastics. Though the ingestion can be found through all most all trophic levels, entanglement is mainly reported with large marine organisms (Li et al., 2018).

Apart from the physical damages, it has been found that surfactants in the digestive tract of biota could cause the rapid release of some chemicals and adsorbed pollutants in MPs into the digestive tract of the organism. These chemicals are fat-soluble, and they can store in fatty tissues of the body and bioaccumulate at higher trophic levels. These chemicals pose a toxic effect on different organisms. Rochman et al. (2014) reported that endocrine-disrupting substances adsorbed to MPs can interfere with the endocrine system of *Oryziaslatipes* (Japanese Medaka), thus resulting in a negative impact on the reproductive capacity of both males and females (Gong and Xie, 2020). PS-particles tend to accumulate in the fish stomach. The accumulation of PS inside oysters can create adverse impacts, such as a reduction in sperm velocity (Li et al., 2018).

Effects of additives in MPs on biota

A variety of additives have been widely used during plastic manufacture to improve physical properties, such as hardness, color, and flame resistance. These additives include nonylphenol, bisphenol A, polybrominated biphenyls ethers, heavy metals, phthalates (potassium acid phthalate), and plasticizers, which are highly toxic to biota. Phthalates and bisphenol can cause endocrine disruption in invertebrates and vertebrates and can also cause cancer and damage to the human reproductive system. During the decomposition of plastics, these additives gradually release into the environment and thus pose serious adverse effects on biota. While many studies focus on the leaching of MPs additives into the environment, there is a lack of studies regarding negative effects of these released additives on biota (Zhang et al., 2019).

Effects of adsorbed toxic substances by MPs on biota

MPs have been reported as a sink for persistent organic pollutants (POPs) and a vector for hydrophobic pollutants, pharmaceuticals, and personal care products (PPCP) in the water (Li et al., 2018; Gong and Xie, 2020). Examples of POPs include polychlorinated biphenyl (PCBs), polycyclic aromatic hydrocarbon (PAHs), and dichlorobiphenyl trichloro-ethane (DDT) combined with heavy metals (e.g., Al, Zn, Pb, Cu, Ag, and Pb). Diethylhexyl phthalate (DEHP), dibutyl phthalate (DBP), diethyl phthalate, diisobutyl phthalate, dimethyl phthalate, benzaldehyde, and 2,4-di-tert- butylphenol have been found in MPs' polymers in the marine environment. Studies have found that adsorbed POPs and heavy metals create worst negative impacts than those released by plastic additives (Zhang et al., 2019). Factors like the degree of weathering, the particle size, and the type of plastic influence the adsorption capacity of MPs (Gong and Xie, 2020). In particular, the coarser structures and higher surface areas of weathered plastics have stronger sorption capacity than virgin plastics. Environmental factors, such as solar irradiation, temperature, microorganisms' activities, and pH, can also affect the sorption desorption process (Zhang et al., 2019). Some studies have shown that MPs can sorb high amounts of PCBs from adjacent water. The higher surface area and polarity of PVC makes more adsorption of heavy metals than PS (Brennecke et al., 2016). Avio et al. (2015) reported higher bioavailability of MPs mixed with PAH after ingestion by mussels (*Mytilusgalloprovincialis*) lead to toxicological consequences. However, the degree of pollutant transfer from MPs to primary consumers and along the food chain is poorly studied and needs to be further researched. While some researchers suggest that human health risks associated with these contaminants are not alarming,others argue that some substances are resulting in acute effects even at small dosages (Seltenrich, 2015).

Furthermore, these adsorb chemical substances on MPs can promote microbial growth and attachment. Ingestion of microbiome-attached MPs can cause inflammation. This also can cause the introduction of pathogenic microorganisms into a clean environment (Zhang et al., 2019). Jiang et al. (2018) reported the presence of potential pathogens in MPs. Barnes and Milner (2005) have found a barnacle (*Eliminiusmodestus*) on plastic debris in the north Atlantic Ocean, which is native to Australia and New Zealand. Biofilms attached to MPs' surfaces can change the physical properties of MPs. The increase in the density of the light MPs can cause the sink of light MPs in the water column and benthic zones. Attached biofilms make the surface of the MPs less hydrophobic. At this time, it is still unknown whether attached biofilms can increase or weaken the adsorption of water-borne pollutants into MPs (Li et al., 2018). These attached microorganisms can enter the food chain and cause negative impacts to higher organisms (Gong and Xie, 2020).

Effects of MPs' biomagnification along the food chain

MPs have been found in a huge number of biota at different trophic levels, indicating transportation along the food chain. MPs have been discovered in aquatic as well as terrestrial animals. Humans can also ingest the MPs through bottled beverages, seafood, and inhalation (Zhang et al., 2019). Humans' intake of MPs through seafood varies between 1to 30 particles per day (Lusher et al., 2017). Van Cauwenberghe and Janssen (2014) show the annual dietary intake of MPs through European shellfish is 11,000 MPs per year. Mussel consumers in the UK would be exposed to upto 123 particles per year. Individuals could be exposed to up to 4,620 particles per year via mussel ingestion in Spain, France, or Belgium (Catarino et al., 2018). *In vivo* studies have demonstrated that MPs can translocate from the gut cavity to the bloodstream and enter cells creating significant adverse effects on the tissue and cellular levels in mussels (Li et al., 2015). However, a recent study conducted by Catarino et al. (2018) provided evidence of human exposure to MPs via ingestion of mussels was lower compared to household fiber exposure during a meal via dust (123–4,620 particles/person per year vs. 13,731–68,415 particles/person per year, respectively). Though there is still no evidence regarding the adverse effects of MPs on humans, there is a high potential for the accumulation of adsorbed pollutants on the surface of MPs inside human body. MPs' toxicity experiments on humans are still at an infancy level, making it hard to accurately decide the adverse effects on human health and thus requiring further studies.

Conclusion

MPs are ubiquitous contaminants present in all environmental compartments including water, sediments, biota, and air. Current evidence on MPs' abundance in aquaculture organisms is currently of high public importance. This chapter summarizes the current status of research that has been carried out related to the detection of MPs in aquaculture organisms, especially the pollution by MPs in mussels. The chapter also addresses in detail the sources of MPs pollution in aquaculture environment, mechanisms of biota-MPs interactions, MPs contamination levels in organisms and the effects of MPs on biota.

Plastics are extensively used by the aquaculture sector from packaging to the manufacturing of various equipment. A considerable portion of equipment and other materials used in aquaculture end up in marine and other aquaculture environments as waste. ALDFG, withering of net cages, plastic lines, and other plastic equipment that are used in aquaculture contribute as direct sources of MPs, while wind, runoff, typhoons, fish meal used as feed in aquaculture, and contaminated baitare identified as indirect sources of MPs entering into the aquaculture environment. These direct and indirect sources contribute to the exposure of aquaculture organisms to MPs. Distribution of MPs has been reported around the globe, including open sea water, coastal areas, inter-tidal zone, estuaries, bays, channels, and ponds, where many types of aquaculture activities take place. This wide distribution raises concern regarding MPs' interactions with biota.

A variety of aquatic organisms, across different trophic levels, can absorb or consume MPs readily. Ingestion and adherence are the two main ways that biota interacts with MPs. Biota can mistake plastics for prey items due to small sizes and indistinguishable nature between food and MPs. Laboratory experiments have studied the interactions of MPs with biota, pathways of MPs within an organism, and accumulation and trophic transfer of MPs. However, laboratory exposure experiments and results should be treated with care as often concentration exceeds more than what is observed in the field.

There are various effects of MPs' particles, additives, and adsorbed substances on biota. Annual dietary intake of MPs through European shellfish can be high as 11,000 MPs per year. These findings show that the consumption of seafood can be a direct way of human exposure to MPs. This urge the necessity of more research regarding the toxicological effects of pollution by MPs on human health, which is still at the infant level.

References

Andréfouët, S., Thomas, Y. and Lo, C. 2014. Amount and type of derelict gear from the declining black pearl oyster aquaculture in Ahe atoll lagoon, French Polynesia. *Marine Pollution Bulletin* 83(1): 224–230. https://doi.org/10.1016/j.marpolbul.2014.03.048

Avio, C.G., Gorbi, S., Milan, M., Benedetti, M., Fattorini, D., D'Errico, G., Pauletto, M., Bargelloni, L., and Regoli, F. 2015. Pollutants bioavailability and toxicological risk from microplastics to marine mussels. *Environmental Pollution* 198: 211–222. https://doi.org/10.1016/j.envpol.2014.12.021

Barnes, D.K.A. and Milner, P. 2005. Drifting plastic and its consequences for sessile organism dispersal in the Atlantic Ocean. *Marine Biology* 146(4): 815–825. https://doi.org/10.1007/s00227-004-1474-8

Bendell, L.I. 2015. Favored use of anti-predator netting (APN) applied for the farming of clams leads to little benefits to industry while increasing nearshore impacts and plastics pollution. *Marine Pollution Bulletin* 91(1): 22–28. https://doi.org/10.1016/j.marpolbul.2014.12.043

Bergmann, M., Gutow, L. and Klages, M. 2015. Marine anthropogenic litter. *In*: *Marine Anthropogenic Litter* (Issue June, pp. 1–447). https://doi.org/10.1007/978-3-319-16510-3

Besseling, E., Wang, B., Lürling, M. and Koelmans, A.A. 2014. Nanoplastic affects growth of S. obliquus and reproduction of D. magna. *Environmental Science and Technology* 48(20): 12336–12343. https://doi.org/10.1021/es503001d

Biernacki, K. 2011. *Application of Plastics in Manufacture* (Issue January). https://www.researchgate.net/publication/304019628_Application_of_Plastics_in_Aquaculture

Bonello, G., Varrella, P. and Pane, L. 2018. First Evaluation of Microplastic Content in Benthic Filter-feeders of the Gulf of La Spezia (Ligurian Sea). *Journal of Aquatic Food Product Technology* 27(3): 284–291. https://doi.org/10.1080/10498850.2018.1427820

Bordós, G., Urbányi, B., Micsinai, A., Kriszt, B., Palotai, Z., Szabó, I., Hantosi, Z. and Szoboszlay, S. 2019. Identification of microplastics in fish ponds and natural freshwater environments of the Carpathian basin, Europe. *Chemosphere* 216: 110–116. https://doi.org/10.1016/j.chemosphere.2018.10.110

Bråte, I.L.N., Eidsvoll, D.P., Steindal, C.C. and Thomas, K.V. 2016. Plastic ingestion by Atlantic cod (Gadus morhua) from the Norwegian coast. *Marine Pollution Bulletin* 112(1–2): 105–110. https://doi.org/10.1016/j.marpolbul.2016.08.034

Brennecke, D., Duarte, B., Paiva, F., Caçador, I. and Canning-Clode, J. 2016. Microplastics as vector for heavy metal contamination from the marine environment. *Estuarine, Coastal and Shelf Science* 178(January): 189–195. https://doi.org/10.1016/j.ecss.2015.12.003

Browne, Mark A., Galloway, T.S. and Thompson, R.C. 2010. Spatial patterns of plastic debris along estuarine shorelines. *Environmental Science and Technology* 44(9): 3404–3409. https://doi.org/10.1021/es903784e

Browne, Mark Anthony, Chapman, M.G., Thompson, R.C., Amaral Zettler, L.A., Jambeck, J. and Mallos, N.J. 2015. Spatial and Temporal Patterns of Stranded Intertidal Marine Debris: Is There a Picture of Global Change? *Environmental Science and Technology* 49(12): 7082–7094. https://doi.org/10.1021/es5060572

Browne, Mark Anthony, Crump, P., Niven, S.J., Teuten, E., Tonkin, A., Galloway, T. and Thompson, R. 2011. Accumulation of microplastic on shorelines woldwide: Sources and sinks. *Environmental Science and Technology* 45(21): 9175–9179. https://doi.org/10.1021/es201811s

Browne, Mark Anthony, Underwood, A.J., Chapman, M.G., Williams, R., Thompson, R.C. and Van Franeker, J.A. 2015. Linking effects of anthropogenic debris to ecological impacts. *Proceedings of the Royal Society B: Biological Sciences* 282(1807). https://doi.org/10.1098/rspb.2014.2929

Castro, R.O., Silva, M.L., Marques, M.R.C. and de Araújo, F.V. 2016. Evaluation of microplastics in Jurujuba Cove, Niterói, RJ, Brazil, an area of mussels farming. *Marine Pollution Bulletin* 110(1): 555–558. https://doi.org/10.1016/j.marpolbul.2016.05.037

Catarino, A.I., Macchia, V., Sanderson, W.G., Thompson, R.C. and Henry, T.B. 2018. Low levels of microplastics (MP) in wild mussels indicate that MP ingestion by humans is minimal compared to exposure via household fibres fallout during a meal. *Environmental Pollution* 237: 675–684. https://doi.org/10.1016/j.envpol.2018.02.069

Cedervall, T., Hansson, L.A., Lard, M., Frohm, B. and Linse, S. 2012. Food chain transport of nanoparticles affects behaviour and fat metabolism in fish. *PLoS ONE* 7(2): 1–6. https://doi.org/10.1371/journal.pone.0032254

Chagnon, C., Thiel, M., Antunes, J., Ferreira, J.L., Sobral, P. and Ory, N.C. 2018. Plastic ingestion and trophic transfer between Easter Island flying fish (Cheilopogon rapanouiensis) and yellowfin tuna (Thunnus albacares) from Rapa Nui (Easter Island). *Environmental Pollution* 243: 127–133. https://doi.org/10.1016/j.envpol.2018.08.042

Cho, Y., Shim, W.J., Jang, M., Han, G.M. and Hong, S.H. 2019. Abundance and characteristics of microplastics in market bivalves from South Korea. *Environmental Pollution* 245: 1107–1116. https://doi.org/10.1016/j.envpol.2018.11.091

Christaki, U., Dolan, J.R., Pelegri, S. and Rassoulzadegan, F. 1991. *I/lln.* 458–464.

Cole, M., Lindeque, P., Fileman, E., Halsband, C. and Galloway, T.S. 2015. The impact of polystyrene microplastics on feeding, function and fecundity in the marine copepod Calanus helgolandicus. *Environmental Science and Technology* 49(2): 1130–1137. https://doi.org/10.1021/es504525u

Cole, M., Lindeque, P., Fileman, E., Halsband, C., Goodhead, R., Moger, J. and Galloway, T. S. (2013). Microplastic ingestion by zooplankton. *Environmental Science and Technology* 47(12): 6646–6655. https://doi.org/10.1021/es400663f

Covernton, G.A. and Cox, K. 2019. Commentary on: Abundance and distribution of microplastics within surface sediments of a key shellfish growing region of Canada. *PLoS ONE* 14(12): 1–7. https://doi.org/10.1371/journal.pone.0225945

Covernton, G., Collicutt, B., Gurney-Smith, H., Pearce, C., Dower, J., Ross, P. and Dudas, S. 2019. Microplastics in bivalves and their habitat in relation to shellfish aquaculture proximity in coastal British Columbia, Canada. *Aquaculture Environment Interactions* 11(October): 357–374. https://doi.org/10.3354/aei00316

Cózar, A., Echevarría, F., González-Gordillo, J.I., Irigoien, X., Úbeda, B., Hernández-León, S., Palma, Á.T., Navarro, S., García-de-Lomas, J., Ruiz, A., Fernández-de-Puelles, M.L. and Duarte, C.M. 2014. Plastic debris in the open ocean. *Proceedings of the National Academy of Sciences of the United States of America* 111(28): 10239–10244. https://doi.org/10.1073/pnas.1314705111

Davidson, K. and Dudas, S.E. 2016. Microplastic Ingestion by Wild and Cultured Manila Clams (Venerupis philippinarum) from Baynes Sound, British Columbia. *Archives of Environmental Contamination and Toxicology* 71(2): 147–156. https://doi.org/10.1007/s00244-016-0286-4

Davidson, T.M. 2012. Boring crustaceans damage polystyrene floats under docks polluting marine waters with microplastic. *Marine Pollution Bulletin* 64(9): 1821–1828. https://doi.org/10.1016/j.marpolbul.2012.06.005

De Witte, B., Devriese, L., Bekaert, K., Hoffman, S., Vandermeersch, G., Cooreman, K. and Robbens, J. 2014. Quality assessment of the blue mussel (Mytilus edulis): Comparison between commercial and wild types. *Marine Pollution Bulletin, 85*(1), 146–155. https://doi.org/10.1016/j.marpolbul.2014.06.006

Devriese, L.I., van der Meulen, M.D., Maes, T., Bekaert, K., Paul-Pont, I., Frère, L., Robbens, J. and Vethaak, A.D. 2015. Microplastic contamination in brown shrimp (Crangon crangon, Linnaeus 1758) from coastal waters of the Southern North Sea and Channel area. *Marine Pollution Bulletin* 98(1–2): 179–187. https://doi.org/10.1016/j.marpolbul.2015.06.051

Digka, N., Tsangaris, C., Torre, M., Anastasopoulou, A. and Zeri, C. 2018. Microplastics in mussels and fish from the Northern Ionian Sea. *Marine Pollution Bulletin* 135(July): 30–40. https://doi.org/10.1016/j.marpolbul.2018.06.063

Dris, R., Gasperi, J., Rocher, V., Saad, M., Renault, N. and Tassin, B. 2015. Microplastic contamination in an urban area: A case study in Greater Paris. *Environmental Chemistry* 12(5): 592–599. https://doi.org/10.1071/EN14167

FAO. 2018. *World Fisheries and Aquaculture*. http://www.fao.org/3/i9540en/i9540en.pdf

FAO. 2020. *The State of World Fisheries and Aquaculture 2020*.

Farrell, P. and Nelson, K. 2013. Trophic level transfer of microplastic: Mytilus edulis (L.) to Carcinus maenas (L.). *Environmental Pollution* 177: 1–3. https://doi.org/10.1016/j.envpol.2013.01.046

Fischer, E.K., Paglialonga, L., Czech, E. and Tamminga, M. 2016. Microplastic pollution in lakes and lake shoreline sediments—A case study on Lake Bolsena and Lake Chiusi (central Italy). *Environmental Pollution* 213: 648–657. https://doi.org/10.1016/j.envpol.2016.03.012

Floerl, O., Sunde, L.M. and Bloecher, N. 2016. Potential environmental risks associated with biofouling management in salmon aquaculture. *Aquaculture Environment Interactions* 8: 407–417. https://doi.org/10.3354/AEI00187

Gall, S.C. and Thompson, R.C. 2015. The impact of debris on marine life. *Marine Pollution Bulletin* 92(1–2): 170–179. https://doi.org/10.1016/j.marpolbul.2014.12.041

Gasperi, J., Dris, R., Mandin, C. and Tassin, B. 2015. First overview of microplastics in indoor and outdoor air. *15th EuCheMS International Conference on Chemistry and the Environment, September*, 2–4. https://www.researchgate.net/publication/281657363%0AFirst

GESAMP. 2016. Sources, fate and effects of microplastics in the marine environment: part 2 of a global assessment. (IMO, FAO/UNESCO-IOC/UNIDO/WMO/IAEA/UN/UNEP/UNDP). *In*: Kershaw, P.J. (Ed.), Rep. Stud. GESAMP No. 90 (96 pp). *Reports and Studies GESAMP, No. 93, 96 P., 93.*

Geyer, R., Jambeck, J.R. and Law, K.L. 2017. Production, use, and fate of all plastics ever made. *Science Advances* 3(7): 19–24. https://doi.org/10.1126/sciadv.1700782

Gong, J. and Xie, P. 2020. Research progress in sources, analytical methods, eco-environmental effects, and control measures of microplastics. *Chemosphere* 254: 126790. https://doi.org/10.1016/j.chemosphere.2020.126790

Graham, E.R. and Thompson, J.T. 2009. Deposit- and suspension-feeding sea cucumbers (Echinodermata) ingest plastic fragments. *Journal of Experimental Marine Biology and Ecology* 368(1): 22–29. https://doi.org/10.1016/j.jembe.2008.09.007

Hahladakis, J. N., Velis, C. A., Weber, R., Iacovidou, E. and Purnell, P. 2018. An overview of chemical additives present in plastics: Migration, release, fate and environmental impact during their use, disposal and recycling. *Journal of Hazardous Materials* 344: 179–199. https://doi.org/10.1016/j.jhazmat.2017.10.014

Hall, N.M., Berry, K.L.E., Rintoul, L. and Hoogenboom, M.O. 2015. Microplastic ingestion by scleractinian corals. *Marine Biology* 162(3): 725–732. https://doi.org/10.1007/s00227-015-2619-7

Huerta Lwanga, E., Gertsen, H., Gooren, H., Peters, P., Salánki, T., Van Der Ploeg, M., Besseling, E., Koelmans, A. A. and Geissen, V. (2016). Microplastics in the Terrestrial Ecosystem: Implications for Lumbricus terrestris (Oligochaeta, Lumbricidae). *Environmental Science and Technology* 50(5): 2685–2691. https://doi.org/10.1021/acs.est.5b05478

Huntington, T. 2019. Marine Litter and Aquaculture Gear—White Paper. In *Aquaculture Stewardship Council* (Issue November).

Iñiguez, M.E., Conesa, J.A. and Fullana, A. 2016. Marine debris occurrence and treatment: A review. *Renewable and Sustainable Energy Reviews* 64: 394–402. https://doi.org/10.1016/j.rser.2016.06.031

Jang, Y.C., Lee, J., Hong, S., Lee, J.S., Shim, W. J. and Song, Y.K. 2014. Sources of plastic marine debris on beaches of Korea: More from the ocean than the land. *Ocean Science Journal* 49(2): 151–162. https://doi.org/10.1007/s12601-014-0015-8

Jang, Y.C., Lee, J., Hong, S., Mok, J.Y., Kim, K.S., Lee, Y.J., Choi, H. W., Kang, H. and Lee, S. 2014. Estimation of the annual flow and stock of marine debris in South Korea for management purposes. *Marine Pollution Bulletin* 86(1–2): 505–511. https://doi.org/10.1016/j.marpolbul.2014.06.021

Jiang, P., Zhao, S., Zhu, L. and Li, D. 2018. Microplastic-associated bacterial assemblages in the intertidal zone of the Yangtze Estuary. *Science of the Total Environment* 624: 48–54. https://doi.org/10.1016/j.scitotenv.2017.12.105

Karami, A., Golieskardi, A., Choo, C.K., Romano, N., Ho, Y. Bin and Salamatinia, B. 2017. A high-performance protocol for extraction of microplastics in fish. *Science of the Total Environment, 578*, 485–494. https://doi.org/10.1016/j.scitotenv.2016.10.213

Karbalaei, S., Hanachi, P., Walker, T.R. and Cole, M. 2018. Occurrence, sources, human health impacts and mitigation of microplastic pollution. *Environmental Science and Pollution Research, 25*(36), 36046–36063. https://doi.org/10.1007/s11356-018-3508-7

Karlsson, T.M., Vethaak, A.D., Almroth, B.C., Ariese, F., van Velzen, M., Hassellöv, M. and Leslie, H.A. 2017. Screening for microplastics in sediment, water, marine invertebrates and fish: Method development and microplastic accumulation. *Marine Pollution Bulletin* 122(1–2): 403–408. https://doi.org/10.1016/j.marpolbul.2017.06.081

Kukulka, T., Proskurowski, G., Morét-Ferguson, S., Meyer, D.W. and Law, K.L. 2012. The effect of wind mixing on the vertical distribution of buoyant plastic debris. *Geophysical Research Letters* 39(7): 1–6. https://doi.org/10.1029/2012GL051116

Lattin, G.L., Moore, C.J., Zellers, A.F., Moore, S.L. and Weisberg, S.B. 2004. A comparison of neustonic plastic and zooplankton at different depths near the southern California shore. *Marine Pollution Bulletin* 49(4): 291–294. https://doi.org/10.1016/j.marpolbul.2004.01.020

Law, K.L., Morét-Ferguson, S.E., Goodwin, D.S., Zettler, E.R., Deforce, E., Kukulka, T. and Proskurowski, G. 2014. Distribution of surface plastic debris in the eastern pacific ocean from an 11-year data set. *Environmental Science and Technology* 48(9): 4732–4738. https://doi.org/10.1021/es4053076

Lee, J., Lee, J.S., Jang, Y.C., Hong, S.Y., Shim, W.J., Song, Y.K., Hong, S.H., Jang, M., Han, G.M., Kang, D. and Hong, S. 2015. Distribution and Size Relationships of Plastic Marine Debris on Beaches in South Korea. *Archives of Environmental Contamination and Toxicology* 69(3): 288–298. https://doi.org/10.1007/s00244-015-0208-x

Leslie, H.A., Brandsma, S.H., van Velzen, M.J.M. and Vethaak, A.D. 2017. Microplastics en route: Field measurements in the Dutch river delta and Amsterdam canals, wastewater treatment plants, North Sea sediments and biota. *Environment International* 101: 133–142. https://doi.org/10.1016/j.envint.2017.01.018

Li, Jiana, Green, C., Reynolds, A., Shi, H. and Rotchell, J.M. 2018. Microplastics in mussels sampled from coastal waters and supermarkets in the United Kingdom. *Environmental Pollution* 241: 35–44. https://doi.org/10.1016/j.envpol.2018.05.038

Li, Jiana, Lusher, A.L., Rotchell, J.M., Deudero, S., Turra, A., Bråte, I.L. N., Sun, C., Shahadat Hossain, M., Li, Q., Kolandhasamy, P. and Shi, H. 2019. Using mussel as a global bioindicator of coastal microplastic pollution. *Environmental Pollution* 244: 522–533. https://doi.org/10.1016/j.envpol.2018.10.032

Li, Jiana, Qu, X., Su, L., Zhang, W., Yang, D., Kolandhasamy, P., Li, D. and Shi, H. 2016. Microplastics in mussels along the coastal waters of China. *Environmental Pollution* 214: 177–184. https://doi.org/10.1016/j.envpol.2016.04.012

Li, Jiana, Yang, D., Li, L., Jabeen, K. and Shi, H. 2015. Microplastics in commercial bivalves from China. *Environmental Pollution* 207: 190–195. https://doi.org/10.1016/j.envpol.2015.09.018

Li, Jingyi, Liu, H. and Paul Chen, J. 2018. Microplastics in freshwater systems: A review on occurrence, environmental effects, and methods for microplastics detection. *Water Research* 137: 362–374. https://doi.org/10.1016/j.watres.2017.12.056

Liboiron, M., Liboiron, F., Wells, E., Richárd, N., Zahara, A., Mather, C., Bradshaw, H. and Murichi, J. 2016. Low plastic ingestion rate in Atlantic cod (Gadus morhua) from Newfoundland destined for human consumption collected through citizen science methods. *Marine Pollution Bulletin* 113(1–2): 428–437. https://doi.org/10.1016/j.marpolbul.2016.10.043

Liu, T.K., Kao, J.C. and Chen, P. 2015. Tragedy of the unwanted commons: Governing the marine debris in Taiwan's oyster farming. *Marine Policy* 53(1): 123–130. https://doi.org/10.1016/j.marpol.2014.12.001

Lusher, A., Hollman, P. and Mandoza-Hill, J.J. 2017. Microplastics in fisheries and aquaculture. In *FAO Fisheries and Aquaculture Technical Paper* (Vol. 615, Issue July). https://doi.org/dmd.105.006999 [pii]\r10.1124/dmd.105.006999

Ma, J., Niu, X., Zhang, D., Lu, L., Ye, X., Deng, W., Li, Y. and Lin, Z. 2020. High levels of microplastic pollution in aquaculture water of fish ponds in the Pearl River Estuary of Guangzhou, China. *Science of The Total Environment* 744: 140679. https://doi.org/10.1016/j.scitotenv.2020.140679

Mathalon, A. and Hill, P. 2014. Microplastic fibers in the intertidal ecosystem surrounding Halifax Harbor, Nova Scotia. *Marine Pollution Bulletin* 81(1): 69–79. https://doi.org/10.1016/j.marpolbul.2014.02.018

Moore, C.J., Moore, S.L., Weisberg, S.B., Lattin, G.L. and Zellers, A.F. 2002. A comparison of neustonic plastic and zooplankton abundance in southern California's coastal waters. *Marine Pollution Bulletin* 44(10): 1035–1038. https://doi.org/10.1016/S0025-326X(02)00150-9

Murphy, C.L. 2018. *A Comparison of Microplastics in Farmed and Wild Shellfish near Vancouver Island and Potential Implications for Contaminant Transfer to Humans By Cassandra Lee Murphy A Thesis Submitted to the Faculty of Social and Applied Sciences in Partial Fulfilment o.* 1–75. https://viurrspace.ca/handle/10613/5540

Murphy, F., Ewins, C., Carbonnier, F. and Quinn, B. 2016. Wastewater Treatment Works (WwTW) as a Source of Microplastics in the Aquatic Environment. *Environmental Science and Technology* 50(11): 5800–5808. https://doi.org/10.1021/acs.est.5b05416

Nelms, S.E., Coombes, C., Foster, L.C., Galloway, T.S., Godley, B.J., Lindeque, P.K. and Witt, M.J. 2017. Marine anthropogenic litter on British beaches: A 10-year nationwide assessment using citizen science data. *Science of the Total Environment* 579: 1399–1409. https://doi.org/10.1016/j.scitotenv.2016.11.137

Nobre, C.R., Santana, M.F.M., Maluf, A., Cortez, F.S., Cesar, A., Pereira, C.D.S. and Turra, A. 2015. Assessment of microplastic toxicity to embryonic development of the sea urchin Lytechinus variegatus (Echinodermata: Echinoidea). *Marine Pollution Bulletin* 92(1–2): 99–104. https://doi.org/10.1016/j.marpolbul.2014.12.050

Paul-Pont, I., Lacroix, C., González Fernández, C., Hégaret, H., Lambert, C., Le Goïc, N., Frère, L., Cassone, A. L., Sussarellu, R., Fabioux, C., Guyomarch, J., Albentosa, M., Huvet, A. and Soudant, P. (2016). Exposure of marine mussels Mytilus spp. to polystyrene microplastics: Toxicity and influence on fluoranthene bioaccumulation. *Environmental Pollution* 216: 724–737. https://doi.org/10.1016/j.envpol.2016.06.039

Pauly, D. and Zeller, D. 2016. Catch reconstructions reveal that global marine fisheries catches are higher than reported and declining. *Nature Communications* 7: 1–9. https://doi.org/10.1038/ncomms10244

Pedà, C., Caccamo, L., Fossi, M.C., Gai, F., Andaloro, F., Genovese, L., Perdichizzi, A., Romeo, T. and Maricchiolo, G. 2016. Intestinal alterations in European sea bass Dicentrarchus labrax (Linnaeus, 1758) exposed to microplastics: Preliminary results. *Environmental Pollution* 212: 251–256. https://doi.org/10.1016/j.envpol.2016.01.083

Peng, L., Fu, D., Qi, H., Lan, C.Q., Yu, H. and Ge, C. 2020. Micro- and nano-plastics in marine environment: Source, distribution and threats—A review. *Science of the Total Environment* 698: 134254. https://doi.org/10.1016/j.scitotenv.2019.134254

Phillips, M.B. and Bonner, T.H. 2015. Occurrence and amount of microplastic ingested by fishes in watersheds of the Gulf of Mexico. *Marine Pollution Bulletin, 100*(1), 264–269. https://doi.org/10.1016/j.marpolbul.2015.08.041

Phuong, N.N., Poirier, L., Pham, Q.T., Lagarde, F. and Zalouk-Vergnoux, A. 2018. Factors influencing the microplastic contamination of bivalves from the French Atlantic coast: Location, season and/or mode of life? *Marine Pollution Bulletin* 129(2): 664–674. https://doi.org/10.1016/j.marpolbul.2017.10.054

Pinheiro, C., Oliveira, U. and Vieira, M. 2017. Occurrence and Impacts of Microplastics in Freshwater Fish. *Journal of Aquaculture & Marine Biology* 5(6). https://doi.org/10.15406/jamb.2017.05.00138

Plastics Europe 2019. *PlasticsEurope, 2019. Plastics—the Facts 2019: An analysis of European plastics production, demand and waste*. https://www.plasticseurope.org/en/resources/market-data

Qu, X., Su, L., Li, H., Liang, M. and Shi, H. 2018. Assessing the relationship between the abundance and properties of microplastics in water and in mussels. *Science of the Total Environment, 621*(February 2019): 679–686. https://doi.org/10.1016/j.scitotenv.2017.11.284

Renzi, M., Guerranti, C. and Blašković, A. 2018. Microplastic contents from maricultured and natural mussels. *Marine Pollution Bulletin* 131(April): 248–251. https://doi.org/10.1016/j.marpolbul.2018.04.035

Rochman, C.M., Kurobe, T., Flores, I. and Teh, S.J. 2014. Early warning signs of endocrine disruption in adult fish from the ingestion of polyethylene with and without sorbed chemical pollutants from the marine environment. *Science of the Total Environment* 493: 656–661. https://doi.org/10.1016/j.scitotenv.2014.06.051

Rodriguez-Seijo, A., Lourenço, J., Rocha-Santos, T.A.P., da Costa, J., Duarte, A.C., Vala, H. and Pereira, R. 2017. Histopathological and molecular effects of microplastics in Eisenia andrei Bouché. *Environmental Pollution* 220: 495–503. https://doi.org/10.1016/j.envpol.2016.09.092

Rummel, C.D., Löder, M.G.J., Fricke, N.F., Lang, T., Griebeler, E.M., Janke, M. and Gerdts, G. 2016. Plastic ingestion by pelagic and demersal fish from the North Sea and Baltic Sea. *Marine Pollution Bulletin* 102(1): 134–141. https://doi.org/10.1016/j.marpolbul.2015.11.043

Savoca, M.S., Wohlfeil, M.E., Ebeler, S.E. and Nevitt, G.A. 2016. Marine plastic debris emits a keystone infochemical for olfactory foraging seabirds. *Science Advances* 2(11): 1–9. https://doi.org/10.1126/sciadv.1600395

Seltenrich, N. 2015. New link in the food chain? Marine plastic pollution and seafood safety. *Environmental Health Perspectives* 123(2): A34–A41. https://doi.org/10.1289/ehp.123-A34

Sussarellu, R., Suquet, M., Thomas, Y., Lambert, C., Fabioux, C., Pernet, M.E.J., Goïc, N. Le, Quillien, V., Mingant, C., Epelboin, Y., Corporeau, C., Guyomarch, J., Robbens, J., Paul-Pont, I., Soudant, P. and Huvet, A. 2016. Oyster reproduction is affected by exposure to polystyrene microplastics. *Proceedings of the National Academy of Sciences of the United States of America* 113(9): 2430–2435. https://doi.org/10.1073/pnas.1519019113

Ta, A. T., Babel, S. and Haarstrick, A. 2020. Microplastics contamination in a high population density area of the chao phraya river, Bangkok. *Journal of Engineering and Technological Sciences* 52(4): 534–545. https://doi.org/10.5614/j.eng.technol.sci.2020.52.4.6

Tanaka, K., Takada, H., Yamashita, R., Mizukawa, K., Fukuwaka, M. aki and Watanuki, Y. 2013. Accumulation of plastic-derived chemicals in tissues of seabirds ingesting marine plastics. *Marine Pollution Bulletin* 69(1–2): 219–222. https://doi.org/10.1016/j.marpolbul.2012.12.010

Taylor, M. L., Gwinnett, C., Robinson, L.F. and Woodall, L.C. 2016. Plastic microfibre ingestion by deep-sea organisms. *Scientific Reports* 6(September): 1–9. https://doi.org/10.1038/srep33997

To, C., Security, F. and All, N.F.O.R. 2016. 2016 In Brief The State of World Fisheries and Aquaculture. *The State of World Fisheries and Aquaculture 2016*, 24.

Ugolini, A., Ungherese, G., Ciofini, M., Lapucci, A. and Camaiti, M. 2013. Microplastic debris in sandhoppers. *Estuarine, Coastal and Shelf Science* 129(June): 19–22. https://doi.org/10.1016/j.ecss.2013.05.026

UNEP. 2016. Marine plastic debris and microplastics. In *Marine Plastic Debris and Microplastics*.

Van Cauwenberghe, L. and Janssen, C.R. 2014. Microplastics in bivalves cultured for human consumption. *Environmental Pollution* 193: 65–70. https://doi.org/10.1016/j.envpol.2014.06.010

Vandermeersch, G., Lourenço, H.M., Alvarez-Muñoz, D., Cunha, S., Diogène, J., Cano-Sancho, G., Sloth, J.J., Kwadijk, C., Damia Barcelo c, I., Allegaert, W., Bekaert, K., Fernandes, J.O., Marques, A., and Robbens, J. 2015. Environmental contaminants of emerging concern in seafood—European database on contaminant levels. *Environmental Research* 143: 29–45. https://doi.org/10.1016/j.envres.2015.06.011

Vandermeersch, G., Van Cauwenberghe, L., Janssen, C.R., Marques, A., Granby, K., Fait, G., Kotterman, M. J.J., Diogène, J., Bekaert, K., Robbens, J. and Devriese, L. 2015. A critical view on microplastic quantification in aquatic organisms. *Environmental Research* 143(2014): 46–55. https://doi.org/10.1016/j.envres.2015.07.016

Wang, J., Lu, L., Wang, M., Jiang, T., Liu, X. and Ru, S. 2019. Typhoons increase the abundance of microplastics in the marine environment and cultured organisms: A case study in Sanggou Bay, China. *Science of the Total Environment* 667: 1–8. https://doi.org/10.1016/j.scitotenv.2019.02.367

Watts, A.J.R., Lewis, C., Goodhead, R.M., Beckett, S.J., Moger, J., Tyler, C.R. and Galloway, T.S. 2014. Uptake and retention of microplastics by the shore crab carcinus maenas. *Environmental Science and Technology* 48(15): 8823–8830. https://doi.org/10.1021/es501090e

Welden, N.A.C. and Cowie, P.R. 2016. Environment and gut morphology influence microplastic retention in langoustine, Nephrops norvegicus. *Environmental Pollution* 214: 859–865. https://doi.org/10.1016/j.envpol.2016.03.067

Wright, S.L., Rowe, D., Thompson, R.C. and Galloway, T.S. 2013. Microplastic ingestion decreases energy reserves in marine worms. *Current Biology* 23(23): R1031–R1033. https://doi.org/10.1016/j.cub.2013.10.068

Zhang, L., Liu, J., Xie, Y., Zhong, S., Yang, B., Lu, D. and Zhong, Q. 2020. Distribution of microplastics in surface water and sediments of Qin river in Beibu Gulf, China. *Science of the Total Environment* 708: 135176. https://doi.org/10.1016/j.scitotenv.2019.135176

Zhang, S., Wang, J., Liu, X., Qu, F., Wang, X., Wang, X., Li, Y. and Sun, Y. 2019. Microplastics in the environment: A review of analytical methods, distribution, and biological effects. *TrAC—Trends in Analytical Chemistry* 111: 62–72. https://doi.org/10.1016/j.trac.2018.12.002

Chapter 8

Occurrence and Source of Microplastics Contamination in Drinking Water and Performance of Water Treatment Plants in Removing Microplastics

Yubraj Dahal and *Sandhya Babel**

INTRODUCTION

Water is an essential part of human life. According to the World Health Organization (WHO), "access to safe drinking water is essential to health, a basic human right and a component of effective policy for health protection" (Water and Organization, 2006). The required intake of water for maintaining health is more than 2.2 liters and 2.3 liters per day for women and men, respectively (Shen et al., 2021). Therefore, it is of utmost importance that drinking water should be free from contaminants.

The quality of drinking water can be maintained through a series of combined actions that include source protection, control of the treatment process, management of the distribution lines, and proper handling of the water system (Water and Organization, 2006). The main purpose of Drinking Water Treatment Plants (DWTPs) is to ensure the quality of water for safe drinking through the elimination of pathogens and undesirable chemicals (Dalmau-Soler et al., 2021). In most countries, the source of drinking water includes both community systems as well as piped water systems (Water and Organization, 2006). Pipeline distribution of drinking water is the most efficient and convenient way of supplying potable water (Water, 2009). Apart from community and piped sources, another common source of drinking water is bottled

School of Bio-chemical Engineering and Technology, Sirindhorn International Institute of Technology, Thammasat University, Khlong Nueng, Khlong Luang, Pathum Thani 12120, Thailand.

*Corresponding author: sandhya@siit.tu.ac.th

water. Bottled water is available in different sizes and volumes in the market, and it offers variability to the consumer's choice and comfort. In 2020, the global market value of bottled water was USD 217.66 billion. This value is estimated to experience a compound annual growth rate (CAGR) of 11.1% from 2021 to 2028 and result in a market value of USD 505.19 billion (Grand View Research, 2021).

However, intriguingly, a new contaminant in drinking water has incited researchers' attention globally. The new contaminant is microplastics (MPs), which are plastic particles less than 5 mm in diameter (Arthur et al., 2009). The first evidence of MPs contamination of the tap water was documented by Kosuth et al. (2018). After this, a series of scientific publications confirmed the presence of the MPs in drinking water around the world. The presence of MPs has been reported in human stools (Schwabl et al., 2019), which confirms human consumption of the MPs. Furthermore, the highest contamination scenario of MPs in bottled and tap water leads to the ingestion of 14,494 and 13,863 MPs particles per day (Shen et al., 2019). Another possible source of MPs' ingestion is the use of MPs' contaminated water for cooking food (Danopoulos et al., 2020). MPs contamination of drinking water is a serious issue owing to the abundance of MPs in various shapes, sizes, and polymeric types. The mixture of MPs contains large amounts of additives that can leach into the human body, such as MPs potentially entering the human tissues upon consumption (Campanale et al., 2020).

The human health risk associated with the MPs contamination of the drinking water is governed by both hazard and level of exposure. The chemicals absorbed by the particles from the environment and the development of the biofilms on the surface of the particles also determine the potential hazards of the MPs (Marsden et al., 2019). Since MPs have been traced in drinking water, polymeric identification of the particles can be correlated with additive types present in the water. The frequently identified polymers—such as polyethylene terephthalates (PET), polyethylene (PE), polypropylene (PP), polyvinyl chloride (PVC), polyamide (PA), polycarbonate (PC), polystyrene (PS), etc.—in the drinking water confirm the presence of toxic additives, such as bisphenol A (BPA), phthalates, styrene monomer, and nonylphenol. In addition, chemical additives such as persistent organic compounds (POPs), polycyclic aromatic hydrocarbons (PAHs), dioxins, and polychlorinated biphenyls (PCBs) are contained in all types of plastics (Alabi et al., 2019). These additives can cause public health impacts, such as mimicking estrogen, causing ovarian disorders, interfering with testosterone and thyroid hormones, alteration in sperm characteristics, inducing possible reproductive and neurological damages as well as developmental and reproductive toxicity, and triggering the formation of deoxy-ribose nucleic acid (DNA) adducts and carcinogenic properties (Halden, 2010). The present level of daily exposure to plastic additives can induce oxidative stress in human cells (Pérez-Albaladejo et al., 2020). Adding to this, MPs that are less than 1.5 μm in size can bypass the gut barrier and reach organs, such as the liver, spleen, and lymph nodes (Danopoulos et al., 2020).

Assessing the MPs contamination of the drinking water is crucial in order to understand the source and level of contamination. This chapter aims to provide a landscape view of MPs contamination of drinking water and demonstrate the efficiency of the treatment plants in removing these MPs particles. The main goals of this chapter are (i) to provide an overview of occurrence (size, morphology, and polymer type) and source of MPs contamination in drinking water that is reported in recent research, (ii) to understand the MPs' removal capacity of different unit processes in DWTPs, and (iii) to know the overall efficacy of DWTPs in removing MPs.

MPs Contamination of Bottled Water

Occurrence of MPs

The sample size, average concentration and range, and size-based concentration of the MPs observed in bottled water are summarized in Table 1. Similarly, Table 2 lists the commonly identified shapes and polymers in bottled water.

Kankanige and Babel (2020b) examined the presence of small-sized MPs in 10 brands of bottled water in Thailand. The particle count in the size range of 6.5–20 μm was nearly seven times higher than the particles that were greater than 50 μm and more than six-fold higher than the particles in the range of 20–50 μm. Furthermore, plastic-bottled water depicted a significantly higher abundance of the MPs compared to glass-bottled water.

In another study, Mason et al. (2018) explored the MPs contamination in bottled water of globally recognized brands. High-ranking brands by volumes, sales, and consumption were chosen from China, India, the USA, Bangladesh, Mexico, the UK, France, Germany, Lebanon, and Indonesia. The particles in the size range (6.5–100 μm) were more than 30 times the concentration that was observed for the size range greater than 100 μm. When considering the brand, the highest average concentration of MPs was observed across Nestle Pure Life and Gerolsteiner at 930 and 807 particles per liter, respectively. In contrast, the lowest concentration was observed at 60.1 and 30 particles per liter in Minalba and San Pellegrino, respectively. Intriguingly, 17 bottled water samples representing seven lots (25% of the total lots) were reported to be free of MPs. In the same manner, a single bottled water sample showed MPs contamination of more than 10,000 particles per liter.

Makhdoumi et al. (2021) investigated the occurrence of MPs in PET bottled mineral water in Iran. MPs were reported in 9 out of 11 brands. Concentration as high as 36 particles per liter was observed in one brand, while the lowest concentration was 2 particles per liter.

Oßmann et al. (2018) studied the small-sized MPs and the pigmented particles in bottled water in Germany. MPs were detected in all the samples with high variability in concentration among different bottle types. Particles that were as high as 35,436 per liter were observed in one of the glass bottles. When comparing the contamination of the new and old reusable PET bottles, old bottles depicted significantly higher (more than three-fold) MPs concentrations than the new bottles. The new reusable plastic bottles contained MPs in the same concentration as seen in the single-use PET bottles.

Zhou et al. (2021) studied MPs contamination of bottled water in China and observed MPs of the size range 100–300 μm in all the bottle samples. Around 90% of the particles were less than 1 mm.

Table 1. Sample Size, Average Concentration and Range, and Size-Based Concentration of the MPs Observed in Bottled Water.

Samples	Average Concentration and Range (Particles Per Liter)	Size-Based Concentration of MPs		References
65 PET bottles 30 Glass bottles	Plastic bottle = 140 ± 19 Glass bottle = 52 ± 4	Size (µm)	Concentrations (Particles Per Liter)	(Kankanige & Babel, 2020b)
		6.5–20	81 ± 3	
		20–50	26 ± 2	
		> 50	12 ± 1	
259 Bottles 11 Brands	Bottled water = 325	Size (µm)	Concentrations (Particles Per Liter)	(Mason et al., 2018)
		6.5–100	315	
		>100	10.4	
24 PET bottles 2 lass bottles 2 PC bottles	Bottled water = 1.9 Range = 0.99–26	25–500 µm (100%)		(Almaiman et al., 2021)
33 PET bottles 11 Brands	Plastic bottle = 8.5 ± 10.2 Range = 0–36			(Makhdoumi et al., 2021)
12 Reusable PET bottles 10 Single use PET bottles 10 Glass bottles	Plastic bottle = 2,649 ± 2,857 Glass bottle = 4,889 ± 5,432	< 5 µm ((95%) < 1.5 µm (50%)		(Oßmann et al., 2018)
30 PET bottles 10 Brands	Plastic bottle = 5.42E7 ± 1.95E7 Range = 3.16E7 – 1.1E8	1.25–4.5 µm (100%)		Zuccarello et al. (2019)
69 PET bottles 23 Brands	Plastic bottle = 2–23 particles per bottle	Size (µm)	Concentration (%)	(Zhou et al., 2021)
		25–50	0.35–14.16	
		50–100	19.39–37.10	
		100–300	29.20–49.20	
		300–500	8.42–15.59	
		500–1,000		
		1,000–5,000	4.75–12.26	

Table 1 contd....

...Table 1 contd.

Samples	Average Concentration and Range (Particles Per Liter)	Size-Based Concentration of MPs					References
11 Brands (single use PET bottles) 15 Brands (reusable PET bottles) 9 Brands (glass bottles) 3 Brands (Beverage cartoon)	Single use (SU) = 14 ± 14, Range = 2–44 Returnable (R) = 118 ± 88 Range = 28–241 Glass bottle (GB) = 50 ± 52 Range = 4–56 Beverage carton (BC) = 11 ± 8 Range = 5–20	Size (μm)	Concentrations (Particles Per Liter)				(Schymanski et al., 2018)
			SU	R	GB	BC	
		5–10	56	41	45	39	
		10–20	29	30	32	28	
		20–50	12	22	14	16	
		50–100	2	5	7	10	
		> 100	1	2	3	7	

Table 2. Morphological (shape) and Polymeric Characteristics of the MPs Observed in Bottled Water.

<table>
<tr><th>Dominant Shape</th><th>Observed Shapes</th><th>Dominant Polymer(s)</th><th colspan="2">Observed Polymers</th><th>References</th></tr>
<tr><td>Fibers</td><td>Fibers = 62.8%
Fragments = 37.2%</td><td>PET</td><td colspan="2">PET, PE, PA, PP</td><td>(Kankanige and Babel, 2020b)</td></tr>
<tr><td>Fragments</td><td>Fiber = 13%
Fragments = 66%
Films = 12%</td><td>PP (54%)</td><td colspan="2">PP (54%), PA (16%), PS (11%), PE (10%), PEST (polyether sulphone (PES) + PET) (6%)</td><td>(Mason et al., 2018)</td></tr>
<tr><td></td><td></td><td></td><td colspan="2">PE, PS, PET</td><td>(Almaiman et al., 2021)</td></tr>
<tr><td>Fragments</td><td>Fibers = 7%
Fragments = 93%</td><td></td><td colspan="2">PET, PS, PP.</td><td>(Makhdoumi et al., 2021)</td></tr>
<tr><td rowspan="2"></td><td rowspan="2"></td><td rowspan="2">PET (plastic bottle)
PE (glass bottle)</td><td>Plastic bottle</td><td>Glass bottle</td><td rowspan="2">(Oßmann et al., 2018)</td></tr>
<tr><td>PET</td><td>PE (46%)
PP (23%)
Styrene-butadiene–copolymer (14%)</td></tr>
<tr><td rowspan="2"></td><td rowspan="2"></td><td></td><td>SU</td><td>R</td><td rowspan="2">(Schymanski et al., 2018)</td></tr>
<tr><td>PEST (SU plastic bottle)
PEST (R plastic bottle)
PEST (glass bottle)
PE (Beverage cartoon)</td><td>PEST = 59%
PE = 9%
PP = 1%
PA = 1%
Others = 30%
Glass bottle
PEST = 41%
PE = 35%
PA = 12%
PP = 8%
Others = 4%</td><td>PEST = 84%
PP = 7%
PE = 5%
PA = 2%
Others = 9%
Beverage
PE = 38%
PEST = 32%
PP = 26%
PA = 1%
Others = 3%</td></tr>
<tr><td>Fibers</td><td>Fibers
(33.3–100%)</td><td>Cellulose (71.16%)</td><td colspan="2">Cellulose (71.16%), PET (6.98%), PE (6.95%), PS (4.65%), PA (4.19%), polyurethane (PU) (1.86%), PVC (1.4%), PP (0.93%), polyacrylic acid (PAA) (1.86%), polyethylene-vinyl-acetate (PEVA) (0.47%), polyacrylamide (PAM) (0.47%)</td><td>(Zhou et al., 2021)</td></tr>
</table>

Schymanski et al. (2018) studied MPs contamination of bottled water in reusable plastic bottles, single-use plastic bottles, and glass bottles in Germany. Surprisingly, returnable bottles had MPs contamination nearly eight and half-folds and 11-folds higher than the single-use bottles and beverage cartons, respectively. The majority of the particles in all packaging types, except returnable bottles, appeared in the 5–20 μm size range. However, returnable bottled water showed more than half of the total MPs in the size range of 5–10 μm. Moreover, the most frequently observed particles in all packing types were sized between 5–10 μm.

Source of MPs

Kankanige and Babel (2020b) indicated that the packaging materials used in bottled water were the main source of MPs contamination. The study further suggests that MPs can be introduced in PET bottled water at various stages, including during water injection into the bottle, storage, transportation, and bottle opening. In addition, the authors further suspect that the MPs that are not screened out by water treatment plants also appear in the bottled water. Other sources include atmospheric emissions during water bottling processes.

Mason et al. (2018) speculate that a notable fraction of the MPs in bottled water is associated with processing activities during the bottling of the water. The authors consider several factors, including particle-fluid dynamics and variability in the manufacturing process as the contributor to MPs. Rather than the water source itself, a significant concentration of the MPs originates from the packaging material. This study suggests abrasion of the cap during the opening is the source of PP in bottled water. In addition, 4% of the MPs were found to be contaminated with industrial lubricants pointing to machinery as the possible source of contamination.

Makhdoumi et al. (2021) consider three possible mechanical mechanisms supporting the release of MPs in bottled water. The first mechanism is the release of MPs from the body of the packaging material due to hydrodynamic shear stress applied during the filling process. Secondly, the mechanical stress produced between the cap and bottle neck during the opening and closing of the bottle results in MPs formation and contamination. Thirdly, MPs can be released to bottled water when the bottle's body is squeezed during transportation.

Oßmann et al. (2018) reported higher concentrations of the pigmented particles (> 10,000 pigmented particles per liter) in reusable bottles that used printed paper as a labeling material compared to those bottles labeled with plastic or single-use glass bottles. Stressing the inner surface of the older bottles during the washing process is one of the reasons for the higher content of PET particles in older bottles. Furthermore, the authors support the theory of abrasion of the cap as the source of MPs contamination in bottled water for both plastic and glass bottled water. However, contamination of glass bottled water by polymers, like PS, styrene-butadiene-copolymer, or PET, cannot be from packaging material alone. The authors speculate that MPs may originate from the abrasion of the machinery parts and contaminate the bottle during cleaning process. In addition, the authors consider that the concentration of the washing liquor and the duration of washing also influence the MPs concentration of the bottled water.

The pH of the bottled water and the density of the material of the bottle showed a strong correlation with the concentration of the MPs. The bottles made up of poor-quality plastics reported higher MPs contamination in bottled water (Zuccarello et al., 2019).

Schymanski et al. (2018) concluded that plastic bottles themselves (packaging material) were the sources of MPs contamination in bottled water. This theory was further supported by the abundance of a high fraction of PEST in returnable bottles. Furthermore, authors suggest abrasion of the packaging material as the source of increased MPs contamination in the returnable bottles because returnable bottles are exposed to high stress due to reutilization as compared to single-use bottles and beverage cartons. The authors reported a correlation between carbon dioxide content and MPs contamination in bottled water. They speculate that the high pressure in the sparkling water compared to the still water increases stress in the packaging material, thereby releasing more particles. Since beverage cartons were coated inside and outside with PE and cellulose material, identification of PE in the water also suggested the origination of the particles from the packaging material itself. Moreover, the authors consider the abrasion of the plastic cap with the harder material of the bottle (neck) to be the plausible reason for extra wear and release of the particles in the bottled water.

Zhou et al. (2021) suggest contamination of the raw water source that is utilized for bottled water production by fibers originating from the washing of synthetic clothes as one of the sources of fibrous MPs. Furthermore, they point to packaging material as the source of MPs in bottled water.

In summary, MPs have been identified in the majority of the bottled water samples. When considering the MPs' size, the MPs' abundance significantly increases with decreasing size. The observed shapes in bottled water are fibers, fragments, and films. The commonly observed polymers are PET, PE, PA, PS, and PP. Though the majority of the MPs originates from the packaging material itself, the level of contamination depends on various stressing factors, such as washing, filling, squeezing while transportation, and abrasion while closing and opening of the bottle cap. Understanding the upper-stress limit that generates minimal MPs from the packaging material while undergoing these processes is imperative to minimize the level of contamination in bottled water. Another possible way of minimizing contamination is shifting from plastic bottled water to glass bottled water; however, it is not an easy paradigm either. Shifting from plastic to glass bottled water production is linked with technical transformation, consumer acceptability, and the manufacturer's economic viability, which can result in a business failure. Similarly, replacing the squeezable plastic bottle with a hard non-squeezable plastic bottle may reduce the contamination caused by external stress.

MPs' Contamination of Tap Water

Occurrence of MPs

Table 3 lists the sample size, average concentration and range, and size-based concentration of the MPs observed in tap water. Table 4 summarizes the commonly identified shapes and polymers in tap water.

Table 3. Sample Size, Average Concentration and Range, and Size-Based Concentration of the MPs observed in Tap Water.

Samples	**Average Concentration and Range**	**Size-Based Concentration of MPs**		**References**
5 Sampling points 45 liter	Tap water 96 particles per liter	Size (μm)	Concentrations (Particles Per Liter)	(Kankanige and Babel, 2020a)
		6.5–53	56 ± 14	
		53–300	21± 7	
		300–500	13 ± 5	
		> 500	6 ± 3	
5 Sampling points	Distribution lines 174 ± 405 per m^3	30% of the MPs less than 20 μm 8–374 μm (size of the MPs)		(Kirstein et al., 2021)
42 Sampling points 42 Samples	Water reserve = 2–1 particles Distribution line = 1 particle Hydrants = 1 particle All concentrations are in per 250 mL			(Paredes et al., 2019)
159 Samples	Tap water 5.45 particles per liter (0–61 particles per liter)	Size range 0.10–5 mm Average size 0.96 mm		(Kosuth et al., 2018)
32 Samples	Tap (South wing) 97 ± 55 per 500 mL (24–228 per 500 mL) Tap (North wing) 219 ± 158 per 500 mL (48–597 per 500 mL)	Size range 6–50 μm		(Pratesi et al., 2021)
42 Samples	Tap water 39 ± 44 particles per liter (1.9–255 particles per liter)	Size (μm)	Concentration (Particles per liter)	(Mukotaka et al., 2021)
		19–50	15 ± 17	
		50–100	17 ± 23	
		> 100	7.3 ± 9.1	

42 Samples 126 L	Public fountain 18 ± 7 particles per liter (5 ± 2–91 ± 14 particles per liter)	Size (mm)	Concentration (%)	(Shruti et al., 2020)
		< 0.5	50	
		0.5–1	25	
		3–5	3	
38 Samples	Tap water 440 particles per liter (0–1,247 particles per liter)	Size (µm)	Concentration (%)	(Tong et al., 2020)
		1–50	31.25–100	
		50–100	1.47–31.25	
		100–300	1.72–31.25	
		300–500	1.18–7.69	
		500–5,000	1.72–11.76	
110 Samples	Tap water 2.181 ± 0.165 particles per liter (0 to 8.605 particles per liter)	Size (µm)	Concentration (%)	(Lam et al., 2020)
		2.7–149	8.9	
		150–499	30.8	
		500–999	25.5	
		1,000–2,499	27.2	
		2,500–5,000	7.6	
21 samples	Tap water 0.7 ± 0.6 particles per liter (0.3–1.6 particles per liter)	Size (µm)	Concentration (%)	(Zhang et al., 2020)
		< 100	1.20	
		100–500	26	
		500–1,000	46.4	
		1,000–5,000	26.4	
4 Samples	Tap water 343.5 particles per liter (267–404 particles per liter)			(Shen et al., 2021)

Table 4. Morphological (Shape) and Polymeric Characteristics of the MPs Observed in Tap Water.

Dominant Shape	Observed Shapes	Dominant Polymer(s)		Observed Polymers	References
		Size (μm)	Polymer		
Fibers	Fibers = 58% Fragments = 37% Films = 5%	300–500	PE (26.5%), PVC (19.1%)	PE, PVC, PP PA, polytetrafluoroethylene (PTFE), PAM	(Kankanige and Babel, 2020a)
		> 500	PE (18.7%), PVC (18.1%)		
Fragments	Fibers = 19% Fragments = 81%	PE (67%)		PES, PE, PP, acrylic	(Kirstein et al., 2021)
Fibers	Fibers = 98% Fragments = 1% Films = 1%				Kosuth et al. (2018)
Fragments	Fragments = 90.4% Fibers = 9% Spheres = 0.6%	PS		PS, PE, styrene ethylene butylene styrene block copolymer (SEBS), PE, PET, PVC, acrylic compounds	(Mukotaka et al., 2021)
Fibers	Fibers, Fragments	PTT		Polytrimethylene terephthalate (PTT), epoxy resin	(Shruti et al., 2020)
Fragments	Fragments 53.85–100% Fibers 1.18–30.77% Spheres 2.27–36.36%	PE (26.8%). PP (24.4%)		PE, PP, copolymerized compounds of PE and PP (22%), PPS (7.3%), PS (6.5%), PET (3.3%)	(Tong et al., 2020)
Fibers	Fibers (98.7%) Films (2.2%)				(Lam et al., 2020)
Fibers	Fibers = 99.2% Fragments = 0.8%	Rayon		Rayon, PET, PE, PS, PAA, polymethyl-phenyl-siloxane (PMPS), polyisoprene (PI), PAM, polydimethylsiloxane (PDMS), polycaprolactone (PCL) – diol	(Zhang et al., 2020)
Fibers	Fibers (50–52%) Fragments (48–50%) Spheres 1%	PE (24%)		PE (24%), PP (4%), PET (25%), PVC, PA	(Shen et al., 2021)

Kankanige and Babel (2020a) examined the MPs' contamination in tap water at Thammasat University in Thailand. The water comes through PVC pipes to the tap. Results showed an increasing concentration of the MPs in smaller size range. The lowest size range, 6.5–53 μm, comprised most of the MPs.

Kosuth et al. (2018) examined the contamination of tap water in 14 countries. The countries taken into consideration for this study are Cuba (n = 1), Ecuador (n = 24), England (n = 3), France (n =1), Germany (n = 2), India (n = 17), Indonesia (n = 21), Ireland (n = 1), Italy (n = 1), Lebanon (n = 16), Slovakia (n = 8), Switzerland (n = 2), Uganda (n = 26), and the US (n = 36). About 81% of the samples were contaminated with anthropogenic debris. Moreover, the highest mean concentration was observed in the US (9.24 particles per liter). In terms of region, North America (the US and Cuba) reported the highest mean concentration of 9.18 particles per liter, while the lowest concentration was observed in European nations (England, France, Germany, Ireland, Italy, Slovakia, and Switzerland) at 3.60 particles per liter. The samples from less developed nations (Cuba, Ecuador, India, Indonesia, and Uganda) depicted low average concentration compared to the developed nations (EU, the US, and Lebanon). The concentration of the particles observed in developed and less developed nations was 6.85 and 4.26 particles per liter, respectively.

Pratesi et al. (2021) examined tap water contamination by MPs in the city of Brasilia, Brazil. Samples (each of 500 ml) were randomly collected from bars, restaurants, and cafes in the South and North Wing of the city. Both areas take water from the same treatment plant that includes unit processes, such as coagulation, flocculation, filtration, and fluoridation. All the samples depicted the presence of the MPs. However, the North Wing reported a significantly higher concentration than the South Wing. The authors speculate the differences in the composition of the water tank, plumbing material (galvanized steel pipes vs. PVC pipes), and sample collection days between these sites to be the reasons for this variability.

Mukotaka et al. (2021) investigated the MPs contamination of tap water in developed countries, including Japan, the US, France, Germany, and Finland. Three to five samples were collected at each site. MPs were found in all the samples. Also, 11 out of 42 samples displayed MPs concentrations were less than 10 particles per liter, whereas three samples displayed MPs were greater than 100 particles per liter. Japan displayed the lowest concentration of MPs (14 particles per liter) than the EU (58 particles per liter). The size of the MPs ranged from 19.2 μm to 4.2 mm. MPs greater than 100 μm were detected in 34 out of 42 samples and in all the sites.

Shruti et al. (2020) studied MPs contamination of free public drinking water fountains in Mexico. All the samples depicted MPs contamination. Nineteen stations out of 42 stations had MPs concentrations of 5–10 particles per liter; 29% of the stations reported MPs count of 13–20 per liter; 22–38 particles per liter were detected in 19% of the stations, and 7% displayed MPs count of 60–91 particles per liter. A higher abundance of MPs was observed in the commercial zones, whereas the lowest was reported in the hospital and recreational zone. MPs less than 1 mm in size accounted for 75% of the total MPs. Only fibers were found to be greater in size than 1 mm, whereas fragments were observed to be smaller than 0.5 mm.

Tong et al. (2020) examined MPs contamination of tap water in different cities in China. Except for two samples, 36 samples depicted MPs contamination. The size of the MPs ranged from 3 to 4,453 μm. Particles less than 50 μm predominated all the samples (abundance ranging from 31.25 to 100%). This study revealed the occurrence of large-sized MPs (> 300 μm) in more than two-thirds of the samples.

Lam et al. (2020) studied tap water samples sourced from surface water in Hong Kong to examine the MPs contamination. Out of 110 samples, 86 samples contained MPs. MPs that were smaller than 1 mm accounted for 97.8% of the total MPs. The size of the MPs varied between 50–4,830 μm.

Zhang et al. (2020) collected tap water samples from seven districts of Qingdao in China for analysis. The majority of the particles were greater than 500 μm. In tap water, 11 polymers were detected. In the study carried out by Shen et al. (2021) in China, researchers found decreasing concentrations of the MPs with increasing size range, and they also discovered MPs greater than 100 μm were minimal. The majority of the particles were smaller than 10 μm.

Source of contamination

PE, PVC, and PA were the dominant polymers in the study carried out by Kankanige and Babel (2020a). In agreement with Mintenig et al. (2019), the authors suggest abrasion of the pipe and fittings that runs from the treatment plant to the household as the source of contamination of these polymers. Furthermore, the authors suggest a higher possibility of contamination of the raw water entering the treatment plants by PET and other polymers. They believe MPs contamination of the raw water (Chao-Phraya River), which has been reported to contain MPs contamination by Ericsson and Johansson (2018), is the source of MPs in the tap water. Another source of contamination was PAM, an anionic polyelectrolyte, which was added to enhance flocculation in the water treatment plant.

Kirstein et al. (2021) agree with Mintenig et al. (2019), who suggest degradation of the equipment (made up of PVC, PE, and PP) used in DWTP and distribution system as the sources of MPs contamination in tap water. The authors found no correlation between the age of the pipe and the concentration of the MPs.

Paredes et al. (2019) speculate that the presence of MPs in the drinking water may be due to the use of plastic [low density polyethylene (LDPE)] coating in the storage tanks. They suggest the hardness of water to be the reason behind significantly less abundance of MPs compared to other studies. The hardness of water develops a layer inside the PVC pipe preventing its wear and tear.

Kosuth et al. (2018) suggest that the difference in concentration of the particles between developed and less developed countries depends on the variability in the water source (well, surface, and snowmelt), regional population density, and water filtration method. Contamination by MPs in treated water may also originate from aerosols and distribution pipes (Mukotaka et al., 2021).

Shruti et al. (2020) suggest a wide spectrum of factors, such as human activities, water transportation means and pathways, and urban water storage conditions, responsible for the presence of MPs in public drinking water fountains. They speculate atmospheric deposition of the MPs at the water intake is the source of fibers in fountain water, along with the discharge of wastewater as the potential source of MPs in the water.

Tong et al. (2020) consider pipes and fittings used in households as the source of MPs, such as PE and PP. Lam et al. (2020) suggest that human activities around the water source, atmospheric deposition of the MPs, and use of plastic equipment in the treatment plant and distribution lines are the potential sources of the MPs.

Zhang et al. (2020) also point to human activities and atmospheric deposition of MPs as the potential source of MPs contamination of the raw water, which eventually appears in the tap water. Furthermore, they point to the degradation of larger MPs during physical and chemical treatment processes in the DWTP as the reason behind the increased concentration of MPs in the tap water compared to the raw water.

To summarize, the highest and lowest concentration reported in tap water was 440 and 0.7 particles per liter, respectively. The observed shapes in tap water are fibers, fragments, spheres, and films. The commonly observed polymers in tap water are PE, PP, PVC, PS, and PA. MPs in the tap water originates mainly from the raw water, and pipes and fittings that run from the treatment plant to the tap. Thus, tap water includes the same contaminants as that of the treatment plants along with the addition of new contaminants from distribution lines. The MPs contamination of tap water can be minimized by increasing the efficacy of the DWTPs and replacing plastic pipes and fittings with metal. An option to prevent MPs consumption from drinking tap water would be the development of portable MPs removal filters that can be installed to the tap.

Performance of Treatment Plants

Occurrence of MPs

Table 5 lists the concentration of the MPs in raw and treated water and the size-based concentration of the MPs observed in DWTPs in different studies. Table 6 summarizes the commonly observed shapes and polymers in the DWTPs. Table 7 shows the removal efficiency of the treatment units and the treatment plants.

Kankanige and Babel (2021) examined the MPs contamination and the removal efficiency of different treatment units in a conventional drinking water treatment plant in Thailand. The prevalence of the MPs was dominated by the small-sized particles ranging in size between 6.5 to 53 and 53 to 300 μm in both the dry and rainy seasons. The major treatment units responsible for the MPs' removal were clarification and dual media filtration. In addition, smaller particles were sedimented and removed effectively compared to the larger particles. Filtration appeared efficient in removing MPs greater than 300 μm, whereas clarification removed all sized particles with better efficiency.

Pivokonsky et al. (2018) investigated the MPs abundance in the raw and treated water of three different treatment plants with varying raw water sources. MPs were observed in all the raw water samples from three different treatment plants. The removal efficiency of the treatment plants varied from 45–100% depending on the size of the MPs. The lower removal rate of 45–51% (depending on the sampling day) was observed at DWTP2 for particle size of 10–50 μm. The authors suggest that the shape of the particles also plays an important role in determining the removal efficiency of treatment plants.

Table 5. Overall and Size-Based Concentration of the MPs in Raw and Treated Water of DWTPs.

	Concentration of MPs (Particles Per Liter)		**Size-Based Concentration (Particles Per Liter)**					**References**
	Raw Water	**Treated Water**						
Dry season	1,385	448.7	Size (μm)	Dry season		Dry season		(Kankanige and Babel, 2021)
				Raw water	Treated water	Raw water	Treated water	
			> 500	161.9	32.1	172.1	59	
Rainy season	1,796.6	769.4	500–300	244.4	51.7	312.6	105.7	
			300–53	472.2	162.4	702.9	281.2	
			53–6.5	506.5	202.5	609	323.5	
	1,473 ± 34 to 3,605 ± 497	338 ± 76 to 628 ± 28	Size (μm)	Raw water		Treated water		(Pivokonsky et al., 2018)
			1–5	40–60		25–60		
			5–10	30–40		30–50		
DWTP1	2,808 ± 80	1,401 ± 86	< 10 μm (65–87%)					(Adib et al., 2021)
DWTP2	1,996 ± 268	1,042 ± 269						
DWTP3	2,172 ± 119	971 ± 103						
	17.88	6.99	Maximum MPs less than 100 μm					(Sarkar et al., 2021)
DWTP	6,614 ± 1,132	930 ± 71	Size (μm)	Raw water		Treated water		(Wang et al., 2020)
			1–5	3,760 ± 726		793 ± 53		
			5–10	1,520 ± 258		136 ± 22		
			10–50	731 ± 216		1 ± 1		
			50–100	379 ± 117		0		
			> 100	224 ± 126		0		

DWTP1	23 ± 2	14 ± 1		Raw water	Treated water	(Pivokonský et al., 2020)
DWTP2	1,296 ± 35	151 ± 4	DWTP1	50% (1–5 µm)	65% (1–5 µm)	
			DWTP2	46% (50–100 µm)	59% (< 50 µm)	
DWTP1	6.6	3.1	Size (µm)	Raw Water	Treated Water	(Radityaningrum et al., 2021)
			1–100	0.5	0.3	
			101–350	4.5	1.3	
			351–1,000	13.5	8.5	
			1000–5000	8.3	2.2	
DWTP2	8.8	2.1	Size (µm)	Raw Water	Treated Water	
			1–100	1.5	0.7	
			101–350	4.9	2.8	
			351–1,000	15.8	3	
			1,000–5,000	12.8	2	
DWTP	0.96 ± 0.46	0.06 ± 0.04		Fibers	Fragments	(Dalmau-Soler et al., 2021)
			Raw Water	57% (200–500 µm)	60 % (500–1,000 µm)	
				15% (20–200 µm)	20% (1,000–2,000 µm)	
				14% (500–1000 µm)	20% (> 2,000 µm)	
				14% (1,000–2,000 µm)	22% (20–200 µm)	
			Treated Water	31% (200–500 µm)	56% (200–500 µm)	
				8% (20–200 µm)	22% (500–1,000 µm)	
				37% (500–1,000 µm)		
				24% (1000–2,000 µm)		
DWTP	2753 (2,173–3,998)	351.9 (336–400)				(Shen et al., 2021)
DWTP	330.2	105.8				(Ferraz et al., 2020)

Table 6. Morphological (Shape) and Polymeric Characteristics of the MPs Observed in DWTPs.

Dominant Shape(s)	Observed Shapes	Dominant Polymer(s)	Observed Polymers	References
Fibers	Dry season Fibers = 49.6% Fragments = 35.1%	PE	PA, PE, PP, PET, PVC	(Kankanige and Babel, 2021)
	Rainy season Fibers = 39.8% Fragments = 38.6%			
Fragments	DWTP1 & DWTP2 Fragments = 71–76%	PET	PET, PE, PP	(Pivokonsky et al., 2018)
	DWTP3 Fibers = 37–61% Fragments = 42–48%			
Fibers	Raw water Fibers = 51.1% Fragments = 35.6% Spheres = 13.3%	PP Raw water (PP = 27.3%) Treated water (PP = 24.8%)	PP, PE, PS, PET, PVC, PA, PU, PC, polybutylene terephthalate (PBT), PTFE, and polyvinylidene fluoride (PVDF)	(Adib et al., 2021)
	Treated water Fibers = 38% Fragments = 56.7% Spheres = 5.3%			
Fibers	Fibers = 52–59% Fragments = 41–48%	Raw water PET (54%) Treated water PET (56%)	PET, PE, PP, PS	(Sarkar et al., 2021)
Fibers	Raw water Fibers = 53.9–73.9% Fragments = 17.6–25.5% Spheres = 8.6–20.6%	PET	PE, PP, PAM, PS, PVC	(Wang et al., 2020)
	Treated water Fibers = 51.6–78.9% Fragments = 14.4–38.3% Spheres = 6.7–10.1%			

Fragments	DWTP1 Fibers = 20% Fragments = 80%				DWTP1 CA (42%)	Cellulose acetate (CA), PET, PVC, PE, PP	(Pivokonský et al., 2020)
	DWTP2 Fibers = 8–13% Fragments = 87–92%				CA, PET, PVC, PE, PP (90%)		
Fibers	DWTP1	Raw (%)		Treated (%)	PE, PP, PE	PE, PP, PE, PA 6, PA 6, 6, polytrimolithicamide	(Radityaningrum et al., 2021)
	Fibers	92.16		100			
	Fragments	6.72					
	Film	1.12					
	DWTP2	Raw (%)		Treated (%)			
	Fibers	94.57		82.35			
	Fragments	5.14		15.29			
	Film	0		1.18			
	Pellet	0.29		1.18			
Fibers		Fibers		Fragments	PP, PET	PP, PET, polyvinyl acetate (PVAC), epoxy resin, PTFE, PA, alkyd resin, polyacrylonitrile (PAN), PS	(Dalmau-Soler et al., 2021)
	Raw Water	59%		41%			
	Treated Water	58%		42%			
Fragments (raw water) Fibers (treated water)		Raw water (%)		Treated water (%)	Raw water PE (26.8%) Treated water PE (24%)	Raw water PE (26.8%), PP (13.2%), PS (16.5%), PET (16.1%), PVC, polymethyl methacrylate (PMMA)	(Shen et al., 2021)
	Fibers	20–23		45–48			
	Fragments	57–65		42–50		Treated water PE (24%), PP (4%), PET (25%), PVC, PA	
	Spheres	12–18		1–2			
Fibers		Fibers (%)	Pellets (%)	Firms (%)			Ferraz et al. (2020)
	Raw Water	89.4	10.5	0.1			
	Treated Water	80.2	19.1	0.7			

Table 7. Removal Efficiency of the Unit Processes and the Treatment Plants.

Unit Process and Efficiency (%)	**Remarks**	**Overall Efficiency (%)**	**References**
Units / Dry season (%) / Rainy season (%) Screen / 6.2 / 7 Clarification / 36.6 / 26.7 Filtration / 34.8 / 30.9 Chlorination tank / 16.4 / 8.9	River water (water source)	67.5 (dry) 57.2 (rainy)	Kankanige and Babel, 2021)
DWTP1 Coagulation, flocculation, and sand filtration DWTP2 & DWTP3 Coagulation, flocculation, sand filtration, and Granular Activated Carbon (GAC) filtration	DWTP1 (Large valley water reservoir) DWTP2 (Small reservoir) DWTP3 (River water) 95% MPs less than 10 µm in raw and treated water; Complete removal of MPs > 50 µm; 2% of MPs > 100 µm in raw water; Significant fraction of MPs < 1 µm in the raw water (111–2,181 particles per liter); Significant reduction of MPs < 1 µm in the treated water	DWTP1 (70) DWTP2 (81) DWTP3 (82)	(Pivokonsky et al., 2018)
Screening, coagulation and flocculation, sand filtration, and disinfection	River water (water source); Nearly complete removal of MPs > 50 µm; Significant removal of fibers than fragments	DWTP1 (50.1) DWTP2 (48.4) DWTP3 (55.2)	(Adib et al., 2021)
Unit Process Coagulation/sedimentation Removed fibers by (50.7–60.6%) Overall efficiency (40.5–50.5%) Sand Filtration Removal efficiency of 29–44.4% Removed 30.9–49.3% (fibers), 23.5–50.9% (fragments) Ozonation GAC Removed 56.8–0.9% of the MPs Removed 73.7–98.5% of 1–5 µm MPs Removed 38–52.1% (fibers), 76.8–86.5% (fragments)	River water (water source); Coagulation/sedimentation almost completely removed particles > 10 µm; Ozonation coupled with GAC increased MPs removal by 17.2–22.2%; 1–5 µm MPs dominated raw (54.6–58%) and treated water (84.4–86.7%); Effective removal of large MPs	82.1–88.6 Fibers (82.9–87.5%); Fragments (73.1–88.9%); Spheres (89.1–92.7%)	(Wang et al., 2020)
Units / Cumulative Removal Efficiency (%) Pulse clarifier / 63 Sand filtration / 85	River water (water source) Significant fraction of MPs < 100 µm in all the samples; Effectiveness of pulse clarifier and sand filtration in removing MPs > 100 µm; Removed fibers and fragments with same level of efficiency	84.6	(Sarkar et al., 2021)

Table 7 contd. ...

...Table 7 contd.

Unit Process and Efficiency (%)			**Remarks**	**Overall Efficiency (%)**	**References**
DWTP2			River water (water source); 70% of MPs < 10 μm; Sedimentation removed to a similar extent; Filtration removed fragments more efficiently than fibers; Ozonation depicted similar removal pattern to filtration GAC significantly removed fragments ≥ 10 to < 50 μm; improved removal of fibers	DWTP1 (40) DWTP2 (88)	(Pivokonský et al., 2020)
Unit Process	Removal Efficiency (%)				
Sedimentation	62 (Overall)				
Filtration	20 (Overall); 34 (fragments ≥ 50 μm); 19 (fragments ≤ 50 μm)				
Ozonation	1 (Overall)				
GAC	5 (Overall)				
Units	DWTP1 (%)	DWTP2 (%)	River water (water source); 350–1,000 μm sized MPs dominated raw (45–50%) and treated water (36–69%); High efficiency of DWTP1 for MPs 1,000–5,000 μm (72%) and 100–350 μm (71%); High efficiency of DWTP2 for MPs 350–1,000 μm (80%) and 1,000–5,000 μm (84%)	DWTP1 (54) DWTP2 (76)	(Radityaningrum et al., 2021)
Aeration	45	47			
Pre-sedimentation	39	37			
Coagulation	17	N/A			
Filtration	26	N/A			
Unit process	Removal efficiency (%)		Sand filtration significantly reduced MPs concentration	93	(Dalmau-Soler et al., 2021)
Sand filtration	78				
GAC	18				
Reverse Osmosis	54				
Aeration, coagulation/ sedimentation, sand filtration, GAC filtration, disinfection				87	(Shen et al., 2021)
				70	Ferraz et al. (2020)

N/A = Not Available

Adib et al. (2021) investigated the MPs contamination of three conventional drinking water treatment plants in Tehran, Iran. In total, 18 samples (45 L) were analyzed for MPs contamination. The DWTPs seemed to be effective in removing larger particles, while smaller particles were not removed owing to the low density. The authors speculate that the higher removal rate of the fibrous MPs may be due to the trapping of the long end of the fibrous MPs by the sand filter. Moreover, the lower removal efficiency of the DWTPs may be due to not using the PAM in the coagulation process. This study suggests that conventional drinking water treatment plants are not capable of removing MPs less than 10 μm in size.

Sarkar et al. (2021) examined the prevalence and removal efficiency of the MPs at different treatment units of a typical drinking water treatment plant in India. In this study, no change in MPs abundance appeared after pre-treatment and coagulation. Pulse clarification significantly reduced the MPs count from 17.88 to 6.99 particles per liter. Surprisingly, the MPs' concentration at the inlet of the sand filter was higher than that at the clarification outlet, increasing from 6.99 to 11.17 particles per liter. However, filtration significantly reduced the MPs count to 2.75 particles per liter.

Wang et al. (2020) examined the MPs' removal efficiency of different treatment units in one of the largest drinking water treatment plants in China. Depending on the size category, the plant removed 72.3–100% of the particles. Coagulation coupled with sedimentation seemed effective in removing fibrous MPs. Similarly, GAC stood efficient in removing small-sized MPs. Surprisingly, the MPs' concentration in the outlet of the ozonation tank increased by 2.8–16%. Basically, the shape and size of the MPs were the features determining the efficiency of the treatment units.

Pivokonsky et al. (2020) studied the MPs contamination (≥ 1 μm) in raw and treated water of two different DWTPs in the Czech Republic. The first treatment plant (DWTP1) receives water from a dam, while the second one (DWTP2) takes water directly from the Uhlava River. For DWTP1, two samples were collected (one at the inlet and another at the outlet). In the case of the second plant (DWTP2), six samples were collected, including inlet, outlet, and four-unit processes (sedimentation, filtration, ozonation, and GAC). MPs were observed in all the samples, but the concentrations were erratic between the treatment plants and along the unit processes. However, MPs comprised less than 0.02% of the total particles present in the water. In addition, more than 70% of the particles were less than 10 μm. As per Pivokonský et al. (2020), the removal efficiency of the DWTP1 for fragments decreased with decreasing size range. When considering the overall treatment efficiency, DWTP2 removed larger particles with better efficiency.

Radilyaningrum et al. (2021) studied MPs' abundance in two DWTPs in Indonesia. At aeration, MPs in the size range of 350–1,000 μm experienced the highest removal efficiency of 68% at DWTP1, while DWTP2 removed the MPs (1–100 μm) with the highest removal efficiency of 93%. In the pre-sedimentation process, DWTP1 removed MPs of 100–350 μm with an efficiency of 73%. However, a removal efficiency of 55% was achieved at DWTP2 for MPs of size range 350–1,000 μm. DWTP1 and DWTP2 removed fragments and fibers effectively during pre-sedimentation, respectively. No fibers were removed during flocculation-sedimentation in DWTP1 and DWTP2. However, 100% and 61% of fragments were removed during flocculation-sedimentation in DWTP1 and DWTP2, respectively. About 31% of the fibers were removed during filtration in DWTP1. DWTP2 could remove 100% of the pellets.

Dalmau-Soler et al. (2021) examined MPs contamination of a DWTP sourced from Llobregat River in Barcelona, Brazil. DWTP almost removed fibers of all the size ranges along the treatment units, while fragments exhibited a decrease in size range along the treatment units.

Ferraz et al. (2020) studied MPs contamination (> 200 μm) of the raw and treated water sourced from the Sinos river in Southern Brazil. Sixteen samples from the river and sixteen samples from the residence that receives treated water were taken into consideration. One sample of raw water showed no MPs contamination after controlled filter correction. Both raw and treated water depicted a similar distribution pattern of the MPs types (shapes). The minimum and maximum MPs count in raw water was 0 and 940 particles per liter, respectively. Similarly, the minimum and maximum concentration of MPs in the treated water was 2 and 459 particles per liter, respectively. Urbanization gradient showed no effect on the abundance of the MPs.

Mintenig et al. (2019) examined MPs contamination of the raw and treated water of five DWTPs, one consumer household (at the water meter and tap water) supplied by each DWTP, and water from three wells in Germany. MPs were not detected in 14 samples out of 24 samples. Five samples depicted MPs contamination of less than one per cubic meter. Similarly, one to three particles were detected in four samples, and a maximum of seven particles were detected in a single sample. The mean concentration of the MPs was observed to be 0.7 particles per cubic meter and concentration ranged between 0–7 particles per cubic meter. MPs were observed in the size range of 50–150 µm. After blank correction, no fibers were found in the samples. All the identified particles were fragments. PET (62%) dominated all the polymers. Other polymers observed were PVC (14%), PA and epoxy resin (9%), and PE (6%).

Source of MPs

Kankanige and Babel (2021), Dalmau-Soler et al. (2021), and Radityaningrum et al. (2021) suggest MPs contamination of the water source is the major source of MPs in raw and treated water. Furthermore, Kankanige and Babel (2021) speculate that the presence of contamination in the treatment plant may be due to MPs originating from the plants themselves. In line with this, Mintenig et al. (2019) suggest abrasion of the storage tanks, pipes, and fittings as the source of MPs, such as epoxy resin, PVC, and PE, in the treated water. Furthermore, despite strict precautions, samples were found to be contaminated with environmental MPs (Kankanige and Babel, 2021). Shen et al. (2021), Adib et al. (2021), and Pivokonsky et al. (2018) suggest human activities near the water source is a potent factor determining the abundance of the MPs in the raw water. The other factors influencing the concentration of the MPs in the raw water are the quality of the water (Adib et al., 2021), the landscape around the water source (Pivokonský et al., 2020), the source of water and ambient environment (Pivokonsky et al., 2018), the weather conditions, such as as precipitation and flood (Adib et al., 2021; Pivokonsky et al., 2018; Shen et al., 2021), and the land use in the vicinity, such as industrial and residential area (Shen et al., 2021).

Similarly, Shen et al. (2021) suspect the degrading of discarded plastics in the water source, wastewater from domestic laundry, personal care products, and other cleansing media are the sources of MPs in raw water. Ferraz et al. (2020) reported discharge of unfiltered sewage from the sink and washing machines and untreated municipal sewage into the river are the major sources of fibrous MPs in the raw water (Sinos river, Brazil). This was further supported by Pivokonský et al. (2020), wherein authors consider a higher MPs count in the raw water of the DWTP2 is possibly due to the effluent of the wastewater treatment plants (WWTPs) entering the water source. Additionally, the study by Shen et al. (2021) speculates the material of the pipes and fittings used in the household is the source of increased concentration of PVC MPs in tap water than in the treated water.

In summary, the DWTPs in operation around the globe cannot remove MPs completely. MPs concentration as high as 1,401 particles per liter and as low as 0.06 particles per liter has been reported in the treated water. The commonly observed shapes in the DWTPs are fibers, fragments, and spheres, while the observed polymers are PA, PE, PP, PET, PVC, and PS. A variability in MPs removal efficiency has been observed among the DWTPs and between the unit processes among the DWTPs around the globe. With current knowledge, the units responsible for removing MPs are coagulation/flocculation and sedimentation, GAC, clarifier, and filtration. However, these units have displayed a mixture of results in removing MPs of various shapes and sizes. More robust and rigorous research is intrinsic to understanding the removal capacity and features of the unit processes. The major source of MPs contamination in DWTPs is raw water. In this regard, source protection plays a key role in diminishing MPs contamination of both DWTPs and tap water. Technological advancement is necessary to bypass the MPs from the raw water before entering the treatment plant. Discarding plastic equipment in treatment plants can also reduce MPs of both DWTPs and tap water to some extent.

Single-Use Plastics and Policies

Accounting for 40% of packaging types, the worldwide production of plastic marked 380 million tons in 2015 (Plastic Europe, 2019). The majority (about 60%) of plastic packaging is used for food and beverages (Groh et al., 2019). The global accumulation of the plastic waste was about 6.3 billion tons in 2015, of which 567 million tons was recycled, 756 millions tons was incinerated, and about 5 billion tons was left in landfills or natural environment (Geyer et al., 2017). With present consumption and waste management practice, it has been estimated that about 12 billion tons of plastics will find their way to landfills and the environment by the end of 2050 (Giacovelli, 2018). About 10% of the plastics produced find their way to the marine environment, and the rest remains in the terrestrial environment in the form of macro-, micro-, and nano-plastics (Mattsson et al., 2015; Allen et al., 2019).

Single-use plastic (SUP) products, made of polyethylene, are used for carrying goods and discarded after one use (Giacovelli, 2018). SUP shares a greater fraction of the general plastic production in many countries around the globe (Chen et al., 2021). According to the UN report, the most common SUP products are (in the order of magnitude) cigarette buds, plastic drinking bottles, plastic bottle caps, food wrappers, plastic grocery bags, plastic lids, straws and stirrers, other types of plastic bags, and foam take-away containers (Giacovelli, 2018). About a trillion SUP bags are used per year (Earth Policy Institute, 2014). In 2014, The regional distribution of single-use plastic production was 26% in North East Asia, 21% in North America, 17% in the Middle East, 16% in Europe, 12% in Asia and the Pacific, 4% in Central and South America, 3% in Former USSR, and 1% in Africa (Giacovelli, 2018). In 2015, the global per capita consumption of plastic material was 45 kg per person. The highest and the lowest per capita consumption in this year was 139 kg per person in North American Free Trade Agreement (NAFTA) region and 16 kg per person in the Middle East and Africa (Plastic Insight, 2016). Plastics, despite their ability to degrade slowly, can degrade to small sizes (5 mm to 1 μm) (da Costa, 2018). Thus, shifting plastics from macro-plastics (MaPs) into MPs, in the long run, is possible (Allen et al., 2019). Concerns are being raised by governments around the world regarding the increasing plastic pollution. More than 60 countries around the globe have implemented bans or levies to combat single-use plastic pollution (Giacovelli,

2018). Different countries have introduced a variety of schemes to beat single-use plastic pollution. Commonly used tools to limit the use of plastic bags are shown in Table 8.

These policies can be adopted at the national, state, or municipal level to combat single-use plastic pollution. Table 9 summarizes the type of policy implemented by different countries at the national level or local level or both levels to deal with plastic bags and Styrofoam products.

When comparing the policy regulations implementation region-wise, European countries have adopted levies to deal with the use of plastic bags, while the rest of the countries are more inclined toward banning as a policy. Many countries have banned the import, production, sale, and use of plastic bags or non-biodegradable plastic bags completely. In contrast, other countries have implemented a ban on plastic/non-biodegradable plastic bags to a specific size range and/or levies on thicker plastics. Countries, namely Ethiopia, Mozambique, Senegal, South Africa, Tanzania (local level), Uganda, and Zimbabwe, have banned plastic bags < 30 μm. Similarly, Canada (Montreal), Argentina, India, Belgium (local level), and Cote d'Ivoire have banned plastic bags < 50 μm. Australia has banned the use of plastic bags < 35 μm in Tasmania and Queensland. Moreover, countries imposing size-based restrictions are Malawi (60 μm), China and Mongolia (25 μm), and Israel and Sri Lanka (20 μm). Apart from regulation policy, some countries are practicing public-private agreements to cope with single-use plastic pollution. Countries like Thailand, Austria, Finland, Germany, Luxembourg, Spain, Sweden, Switzerland, Canada, and Australia are a few adopting this strategy (Giacovelli, 2018).

Six Southeast Asian countries are in the top 20 countries with unmanaged plastic waste. These are Indonesia (second), Philippines (third), Vietnam (fourth), Thailand

Table 8. Policy Tools to Limit the Use of Plastic Bags.

Policy		**Features**
Regulatory Instruments	Ban	A total or partial restriction (a particular type or combination) on single-use plastics including plastic bags, foamed plastic products, etc. An example of this policy is a ban on the use of plastic bags 30 μm thickness.
Economic Instruments	Levy on suppliers	Under this policy, the suppliers of plastic bags (domestic producers or importers) are liable for paying the levy. To ensure consumers' behavioral change toward the use of plastic bags, this levy should be fully passed on from the suppliers to the retailers given that the retailers charge consumers for plastic bags or promote the use of recyclable bags by rewarding those who obviate the plastic bags.
	Levy on retailers	This is the levy to be paid by the retailer on the purchase of plastic bags. Under this scheme, the retailers are not bound to impose/pass the tax on the consumers.
	Levy on consumers	Under this scheme, the consumers are charged a certain amount as defined by the law on each bag sold.
Combination of regulatory and economic instruments	Ban and Tax	This policy includes a combination of ban and tax. An example of this policy is a ban on thin plastic bags and a tax on thicker bags.

Source: (Giacovelli, 2018)

Table 9. Policies Adopted by Different Countries to Combat With the Plastic Bags and Styrofoam Products at the National Level, Local Level or Both.

Region	Country			
	Ban Only	**Both Levy and Ban or Levy Only**	**Imminent Action**	
			Ban	**Levy**
Africa	Benin (N), Burkina Faso (N), Cameroon (N), Cape Verde (N), Chad (L), Cote d'Ivoire (N), East Africa (R), Egypt (L), Eritrea (N), Ethiopia (N), Gambia (N), Guinea-Bissau (N), Kenya (N), Malawi (N), Mali (N), Mauritania (N), Mauritius (N), Morocco (N), Mozambique (N), Niger (N), Rwanda (N), Senegal (N), Somalia (L), Tanzania (N&L), Uganda (N)	Botswana (N), Tunisia (N), Zimbabwe (N), South Africa (N)	Botswana (N), Nigeria (N), Republic of Congo (N)	
Asia	Bangladesh (N), Bhutan (N), China (L), India (N&L), Indonesia (L), Malaysia (L), Mongolia (N), Myanmar (L), Pakistan (L), Philippines (L), Sri-Lanka (N)	China (N and L), Indonesia (L), Israel (N), Malaysia (L), Vietnam (N)		
Central and South America	Antigua and Barbuda (N), Argentina (L), Belize (N), Brazil (L), Chile (N), Ecuador (L), Guatemala (L), Guyana (N), Haiti (N), Hondurans (L), Mexico (L), Panama (N), St. Vincent and the Grenadines (N)	Brazil (L), Colombia (N), Mexico (L)	Costa Rica (N), Jamaica (N)	Uruguay (N), Croatia (N)
Europe	Belgium (L), France (N), Italy (N), Romania (N), Spain (L)	Belgium (N), Bulgaria (N), Croatia (N), Cyprus (N), Czech Republic (N), Denmark (N), Estonia (N), Greece (N), Hungary (N), Ireland (N), Italy (N), Latvia (N), Lithuania (N), Malta (N), Netherlands (N), Portugal (N), Romania (N), Slovakia (N), Spain (L), Sweden (L), the United Kingdom (L)	Slovenia (N)	Spain (N), Poland (N)
North America	Canada (L), The USA (L)	The USA (L)		
Oceania	Australia (L), Papua New Guinea (N), Vanuatu (N), Marshall Islands (N), Palau (N)	Fiji (N)		Vanuatu (N), New Zealand (L)

Source: (Giacovelli, 2018)
N = National level
R = Regional level
L = Local level
N and L = National and Local Level

Table 10. Legal Framework and Policies Adopted by the Southeast Asian Countries to Tackle Plastic Bags.

ASEAN Countries	Legal Framework and Policies
Brunei Darussalam	Ban plastic bags in all supermarkets by 2019.
Cambodia	Ban on the import, local production, distribution, and use of plastic bags less than 30 μm thickness and base width of 25 cm. Charging of USD 0.10 for one plastic bag by the supermarkets.
Indonesia	Ministry of Industry encouraging the use of biodegradable plastics. Ban on plastic bags in the cities like Banjarmasin and Bogor. Charging of USD 0.01 on plastics bag in 23 cities for a trial period of 3 months (2016).
Lao PDR	Use of recyclable plastic bags in cafes and markets.
Malaysia	Implementation of tax on plastic. Ban on the use of plastic bags and containers made from polystyrene in Selangor and Federal states.
Myanmar	Ban on plastic bags in Mandalay, Nay Pyi Taw, and Yangon city. Charging of USD 0.07 for one single-use plastic bag. Encouragement for carrying own bag.
Philippines	Prohibition on the disposal of plastic products by the Department of Environment and Natural Resources. Legal decree for plastic bag reduction in Quezon City (2012). Proposed actions like the plastic straw and stirrer ban (2018), the total plastic ban act (2011), and national standards on plastic shopping bags and biodegradable plastics. Ban on single-use plastic in Dipolog and Cebu City. Awareness campaign for the use of reusable bags with the slogan "Bring Your Own Bag".
Singapore	Packaging agreement aimed to reduce packaging waste (2007).
Thailand	Implementation of the "Say No to Plastic Bag" campaign in 11,000 mini markets and stores. Signing of Memorandum of Understanding (MOU) between the Ministry of Natural Resources and Environment. Campaign on "No plastic cap seals of drinking water bottles" (2018). Ban on plastic bags in national parks by the Department of National Parks, Wildlife, and Plant Conservation (2018). Ban on plastic bags in hospitals by Department of Medical Services, Ministry of Public Health (2018).Campaign for reducing single-use plastic at campuses: "Public-private partnership for sustainable plastic and waste management" (2018).
Vietnam	Tax on plastic bags ranges from USD 1.3 – USD 2.1 per kilogram of plastic bags. Announcement of the circular for eco-friendly plastic bags.

Source: (Hisham and Florent, 2019)

(sixth), Malaysia (eighth), and Myanmar (seventeenth) (The Conversation, 2021). In terms of Southeast Asian countries, the United Nations Economic and Social Commission for Asia and the Pacific (ESCAP), and Japan have collaboratively launched a project named "Closing the Loop" in four Southeast Asian Nations (ASEAN) cities. These are Kuala Lumpur (Malaysia), Surabaya (Indonesia), Nakhon Si Thammarat (Thailand), and Da Nang (Vietnam), and it was to deal with plastic waste pollution (ESCAP, 2020). Recently, ASEAN nations launched ASEAN Regional Action Plan for Combating Marine Debris in the ASEAN Members States (2021-2025) on May 28, 2021, to tackle plastic pollution (The World Bank, 2021). The action plan binds the countries to undertake actions to eradicate the use of single-use plastics, develop common policies on recycling and plastic packaging standards, and reinforce measurements and monitor mechanisms of marine debris at the regional level (ADB, 2021). Table 10 summarizes the measures adopted by Southeast Asian countries to cope with plastic bags.

Conclusions

MPs have been reported in the majority of the bottled water samples. Studies so far have shown an increasing concentration of MPs with decreasing size range. In terms of packaging material, old reusable plastic bottles have a higher concentration of particles than single-use plastic bottles and glass bottles. Thus, avoiding reutilization of the plastic bottles can be a better option to avoid MPs contamination to some extent. The commonly observed shapes are fibers and fragments. Similarly, the frequently observed polymers are PET, PE, PA, PS, and PP in bottled water. Furthermore, the abundance of polymers other than PET and PP suggests the source of origination of MPs beyond packaging material, such as machinery, bottle-washing liquor, and raw water. The factors that influence the release of the MPs from the bottled water are stressing of the bottles during the filling and transportation processes, abrasion of the bottle cap during opening and closing, density of the packaging material, and pH and carbon dioxide content of the water.

Also, the tap water as well as the systems, such as distribution lines, hydrants, pumping stations, and water reserves, contain MPs. The tap water in the developed countries depicted a higher concentration of MPs compared to less developed countries. This remains a subject of profound research and analysis. The common polymers in tap water are PE, PP, PVC, PS, and PA. The dominant shapes in tap water are fibers and fragments. The primary source of MPs in tap water is the raw water itself. MPs also originate from the degradation of the equipment made up of plastic materials (such as PVC, PE, and PP) that are used in the treatment plants, pipes, and fittings that run from the treatment plants to taps.

The MPs removal efficiency of the DWTPs ranged between 40–93%. This implies that a significant proportion of MPs can be removed by DWTPs. Based on recent studies, it can be concluded that MPs that are less than 10 μm greatly contribute to the MPs concentrations in the raw and treated water. At present, based on the studies so far, it can be concluded that DWTPs cannot remove MPs that are smaller than 10 μm. Thus, suitable technological innovation is needed to curb the MPs from escaping the treatment plant. Moreover, both fibers and fragments shared a major fraction of the MPs in DWTPs. In terms of shape, treatment plants seem equally efficient in removing fibers and fragments. The major unit processes that are responsible for the removal of a significant proportion of MPs are sedimentation, clarification, and filtration. The major source of MPs contamination in the DWTPs is the raw water. The source of raw water (surface and groundwater), the quality of the raw water, the nature of the source (protected areas/open areas), the weather and ambient environmental conditions, and the human activities in the premises of water sources are the factors influencing the abundance of the MPs in the raw water. In addition, MPs also originate from the plastic equipment used in the treatment plant.

With proliferating plastic production globally and consequent environmental implications introduced by their mismanagement, nations around the globe are finding ways to combat plastic pollution, especially packaging types. The commonly employed strategies to cope with plastic pollution are ban and levy on plastic products to a certain specification. Levy policies have been widely employed in European countries, while many other countries have adopted ban policies.

References

ADB. 2021. *Southeast Asia Takes Action against Plastic Pollution.* https://seads.adb.org/news/southeast-asia-takes-action-against-plastic-pollution.

Adib, D., Mafigholami, R. and Tabeshkia, H. 2021. Identification of Microplastics in Conventional.

Drinking Water Treatment Plants in Tehran, Iran. *Research Square.* https://doi.org/10.21203/rs.3.rs-242504/v1.

Alabi, O.A., I, O.K., Awosolu, O. and Alalade, O.E. 2019. Public and Environmental Health Effects of Plastic Wastes Disposal: A Review. *Journal of Toxicology and Risk Assessment* 5(2). https://doi.org/10.23937/2572-4061.1510021

Allen, S., Allen, D., Phoenix, V.R., Le Roux, G., Durántez Jiménez, P., Simonneau, A., Binet, S. and Galop, D. 2019. Atmospheric transport and deposition of microplastics in a remote mountain catchment. *Nature Geoscience* 12(5): 339–344. https://doi.org/10.1038/s41561-019-0335-5

Almaiman, L., Aljomah, A., Bineid, M., Aljeldah, F.M., Aldawsari, F., Liebmann, B., Lomako, I., Sexlinger, K. and Alarfaj, R. 2021. The occurrence and dietary intake related to the presence of microplastics in drinking water in Saudi Arabia. *Environmental Monitoring and Assessment* 193(7): 1–13.

Arthur, C., Baker, J.E. and Bamford, H.A. 2009. Proceedings of the International Research Workshop on the Occurrence, Effects, and Fate of Microplastic Marine Debris, September 9–11, 2008, University of Washington Tacoma, Tacoma, WA, USA.

Campanale, C., Massarelli, C., Savino, I., Locaputo, V. and Uricchio, V.F. 2020. A Detailed Review Study on Potential Effects of Microplastics and Additives of Concern on Human Health. *Int. J. Environ. Res. Public Health* 17(4). https://doi.org/10.3390/ijerph17041212

Chen, Y., Awasthi, A.K., Wei, F., Tan, Q. and Li, J. 2021. Single-use plastics: Production, usage, disposal, and adverse impacts. *Sci Total Environ* 752: 141772. https://doi.org/10.1016/j.scitotenv.2020.141772

da Costa, J.P. 2018. Micro-and nanoplastics in the environment: research and policymaking. *Current Opinion in Environmental Science & Health* 1: 12–16.

Dalmau-Soler, J., Ballesteros-Cano, R., Boleda, M.R., Paraira, M., Ferrer, N. and Lacorte, S. 2021. Microplastics from headwaters to tap water: occurrence and removal in a drinking water treatment plant in Barcelona Metropolitan area (Catalonia, NE Spain). *Environ. Sci. Pollut. Res. Int.* https://doi.org/10.1007/s11356-021-13220-1

Danopoulos, E., Twiddy, M. and Rotchell, J.M. 2020. Microplastic contamination of drinking water: A systematic review. *PLoS One* 15(7): e0236838. https://doi.org/10.1371/journal.pone.0236838

Earth Policy Institute. 2014. *Plastic Bags Fact Sheets* http://www.earth-policy.org/mobile/releases/plastic_bags_fact_sheet

Ericsson, E. and Johansson, E. 2018. Quantification for the Flow of Microplastic Particles in Urban Environment: A Case of the Chao Phraya River, Bangkok, Thailand. *In*: Degree Project in Technology, School of Architecture and the Built ....

ESCAP. 2020. *New UN initiative to reduce plastic pollution from ASEAN cities* https://www.unescap.org/news/new-un-initiative-reduce-plastic-pollution-asean-cities

Ferraz, M., Bauer, A.L., Valiati, V.H. and Schulz, U.H. 2020. Microplastic Concentrations in Raw and Drinking Water in the Sinos River, Southern Brazil. *Water* 12(11). https://doi.org/10.3390/w12113115

Geyer, R., Jambeck, J.R. and Law, K.L. 2017. Production, use, and fate of all plastics ever made. *Science Advance* 3(7). https://doi.org/DOI: 10.1126/sciadv.1700782

Giacovelli, C. 2018. Single-Use Plastics: A Roadmap for Sustainability (rev. 2).

Grand View Research. 2021. *Bottled Water Market Size, Share & Trends Report 2021–2028.* https://www.grandviewresearch.com/industry-analysis/bottled-water-market

Groh, K.J., Backhaus, T., Carney-Almroth, B., Geueke, B., Inostroza, P.A., Lennquist, A., Leslie, H.A., Maffini, M., Slunge, D., Trasande, L., Warhurst, A.M. and Muncke, J. 2019. Overview of known plastic packaging-associated chemicals and their hazards. *Sci Total Environ* 651(Pt 2): 3253–3268. https://doi.org/10.1016/j.scitotenv.2018.10.015

Halden, R.U. 2010. Plastics and health risks. *Annu. Rev. Public. Health.* 31: 179–194. https://doi.org/10.1146/annurev.publhealth.012809.103714

Hisham, M.M. and Florent, M.Z. 2019. Overview of plastic issues in ASEAN, focusing on marine debris and microplastics in the region.

Kankanige, D. and Babel, S. 2020a. Identification of Micro-plastics (MPs) in Conventional Tap Water Sourced from Thailand. *Journal of Engineering and Technological Sciences* 52(1). https://doi.org/10.5614/j.eng.technol.sci.2020.52.1.7

Kankanige, D. and Babel, S. 2020b. Smaller-sized micro-plastics (MPs) contamination in single-use PET-bottled water in Thailand. *Sci. Total Environ.* 717: 137232. https://doi.org/10.1016/j.scitotenv.2020.137232

Kankanige, D. and Babel, S. 2021. Contamination by ≥ 6.5 μm-sized microplastics and their removability in a conventional water treatment plant (WTP) in Thailand. *Journal of Water Process Engineering 40*. https://doi.org/10.1016/j.jwpe.2020.101765

Kirstein, I.V., Hensel, F., Gomiero, A., Iordachescu, L., Vianello, A., Wittgren, H.B. and Vollertsen, J. 2021. Drinking plastics? - Quantification and qualification of microplastics in drinking water distribution systems by microFTIR and Py-GCMS. *Water Res* 188: 116519. https://doi.org/10.1016/j.watres.2020.116519

Kosuth, M., Mason, S.A. and Wattenberg, E.V. 2018. Anthropogenic contamination of tap water, beer, and sea salt. *PLoS One* 13(4): e0194970. https://doi.org/10.1371/journal.pone.0194970

Lam, T.W.L., Ho, H.T., Ma, A.T.H. and Fok, L. 2020. Microplastic Contamination of Surface Water-Sourced Tap Water in Hong Kong—A Preliminary Study. *Applied Sciences* 10(10). https://doi.org/10.3390/app10103463

Makhdoumi, P., Amin, A.A., Karimi, H., Pirsaheb, M., Kim, H. and Hossini, H. 2021. Occurrence of microplastic particles in the most popular Iranian bottled mineral water brands and an assessment of human exposure. *Journal of Water Process Engineering* 39: 101708.

Marsden, P., Koelmans, A., Bourdon-Lacombe, J., Gouin, T., D'Anglada, L., Cunliffe, D., Jarvis, P., Fawell, J. and De France, J. 2019. *Microplastics in Drinking Water* (9241516194).

Mason, S.A., Welch, V.G. and Neratko, J. 2018. Synthetic Polymer Contamination in Bottled Water. *Front Chem* 6: 407. https://doi.org/10.3389/fchem.2018.00407

Mintenig, S.M., Loder, M.G.J., Primpke, S. and Gerdts, G. 2019. Low numbers of microplastics detected in drinking water from ground water sources. *Sci. Total Environ.* 648: 631–635. https://doi.org/10.1016/j.scitotenv.2018.08.178

Mukotaka, A., Kataoka, T. and Nihei, Y. 2021. Rapid analytical method for characterization and quantification of microplastics in tap water using a Fourier-transform infrared microscope. *Science of the Total Environment* 790. https://doi.org/10.1016/j.scitotenv.2021.148231

Oßmann, B.E., Sarau, G., Holtmannspotter, H., Pischetsrieder, M., Christiansen, S.H. and Dicke, W. 2018. Small-sized microplastics and pigmented particles in bottled mineral water. *Water Res*, *141*, 307-316. https://doi.org/10.1016/j.watres.2018.05.027

Paredes, M., Bodero, E., Castillo, T., Fuentes, G. and Viteri, R. 2019. Microplastics in the drinking water of the Riobamba city, Ecuador. *Przegląd Naukowy Inżynieria i Kształtowanie Środowiska* 28(4): 653–663. https://doi.org/10.22630/pniks.2019.28.4.59

Pérez-Albaladejo, E., Solé, M. and Porte, C. 2020. Plastics and plastic additives as inducers of oxidative stress. *Current Opinion in Toxicology*.

Pivokonsky, M., Cermakova, L., Novotna, K., Peer, P., Cajthaml, T. and Janda, V. 2018. Occurrence of microplastics in raw and treated drinking water. *Sci. Total Environ.* 643: 1644–1651. https://doi.org/10.1016/j.scitotenv.2018.08.102

Pivokonský, M., Pivokonská, L., Novotná, K., Čermáková, L. and Klimtová, M. 2020. Occurrence and fate of microplastics at two different drinking water treatment plants within a river catchment. *Science of the Total Environment* 741: 140236.

Plastic Europe. 2019. *Plastics-the Facts 2019 An analysis of European plastics production, demand and waste data* (Available on the website: http://www. plasticseurope. org, Issue.

Plastic Insight. 2016. *Global Consumption of Plastic Materials By Region (1980–2015)*. https://www.plasticsinsight.com/global-consumption-plastic-materials-region-1980–2015/

Pratesi, C.B., AL Santos Almeida, M.A., Cutrim Paz, G.S., Ramos Teotonio, M.H., Gandolfi, L., Pratesi, R., Hecht, M. and Zandonadi, R.P. 2021. Presence and Quantification of Microplastic in Urban Tap Water: A Pre-Screening in Brasilia, Brazil. *Sustainability* 13(11): 6404.

Radityaningrum, A.D., Trihadiningrum, Y., Mar'atusholihah, Soedjono, E.S. and Herumurti, W. 2021. Microplastic contamination in water supply and the removal efficiencies of the treatment plants: A case of Surabaya City, Indonesia. *Journal of Water Process Engineering* 43. https://doi.org/10.1016/j.jwpe.2021.102195

Sarkar, D.J., Sarkar, S.D., Das, B.K., Praharaj, J.K., Mahajan, D.K., Purokait, B., Mohanty, T.R., Mohanty, D., Gogoi, P. and Kumar, S. (2021). Microplastics removal efficiency of drinking water treatment plant with pulse clarifier. *Journal of Hazardous Materials* 413: 125347.

Schwabl, P., Koppel, S., Konigshofer, P., Bucsics, T., Trauner, M., Reiberger, T. and Liebmann, B. 2019. Detection of Various Microplastics in Human Stool: A Prospective Case Series. *Ann. Intern. Med.* 171(7): 453–457. https://doi.org/10.7326/M19-0618

Schymanski, D., Goldbeck, C., Humpf, H.U. and Furst, P. 2018. Analysis of microplastics in water by micro-Raman spectroscopy: Release of plastic particles from different packaging into mineral water. *Water Res.* 129: 154–162. https://doi.org/10.1016/j.watres.2017.11.011

Shen, M., Zeng, Z., Wen, X., Ren, X., Zeng, G., Zhang, Y. and Xiao, R. 2021. Presence of microplastics in drinking water from freshwater sources: the investigation in Changsha, China. *Environ. Sci. Pollut. Res. Int.* https://doi.org/10.1007/s11356-021-13769-x

Shruti, V.C., Perez-Guevara, F. and Kutralam-Muniasamy, G. 2020. Metro station free drinking water fountain- A potential "microplastics hotspot" for human consumption. *Environ. Pollut.* 261: 114227. https://doi.org/10.1016/j.envpol.2020.114227

The Conversation. 2021. *Why collaboration in the ASEAN region is vital to tackle plastic waste in the oceans.* https://theconversation.com/why-collaboration-in-the-asean-region-is-vital-to-tackle-plastic-waste-in-the-oceans-151849

The World Bank. 2021. *ASEAN Member States Adopt Regional Action Plan to Tackle Plastic Pollution* https://www.worldbank.org/en/news/press-release/2021/05/28/asean-member-states-adopt-regional-action-plan-to-tackle-plastic-pollution

Tong, H., Jiang, Q., Hu, X. and Zhong, X. 2020. Occurrence and identification of microplastics in tap water from China. *Chemosphere* 252: 126493. https://doi.org/10.1016/j.chemosphere.2020.126493

Wang, Z., Lin, T. and Chen, W. 2020. Occurrence and removal of microplastics in an advanced drinking water treatment plant (ADWTP). *Sci. Total Environ.* 700: 134520. https://doi.org/10.1016/j.scitotenv.2019.134520

Water, S. and Organization, W.H. 2006. Guidelines for drinking-water quality [electronic resource]: incorporating first addendum. Vol. 1, Recommendations.

Water, U. 2009. Water in a Changing World, United Nations World Water Development Report 3. *World Water Assessment Programme*.

Zhang, M., Li, J., Ding, H., Ding, J., Jiang, F., Ding, N.X. and Sun, C. 2020. Distribution characteristics and influencing factors of microplastics in urban tap water and water sources in Qingdao, China. *Analytical Letters* 53(8): 1312–1327.

Zhou, X.-j., Wang, J., Li, H.-y., Zhang, H.-m., Hua, J. and Zhang, D.L. 2021. Microplastic pollution of bottled water in China. *Journal of Water Process Engineering* 40. https://doi.org/10.1016/j.jwpe.2020.101884

Zuccarello, P., Ferrante, M., Cristaldi, A., Copat, C., Grasso, A., Sangregorio, D., Fiore, M. and Conti, G.O. 2019. Exposure to microplastics (< 10 μm) associated to plastic bottles mineral water consumption: The first quantitative study. *Water Research* 157: 365–371.

Chapter 9

Regulating the Export of Plastic Waste in Canada

Ryder LeBlanc[1,*] and *Gail Krantzberg*[2]

INTRODUCTION

Plastic waste is a serious threat to global health and the environment. Globally, an estimated 8,300 million metric tons of plastic have been produced by humankind (Geyer et al., 2017). This total estimation increases by approximately 300 million tons each year (IUCN, 2021). It is estimated that Canada contributes 3.3 million tonnes of plastic waste annually (Young, 2021). In terms of final disposal for these products, 9% is recycled, 12% is incinerated, and the remaining 79% is left discarded in landfills, bodies of water, or other areas of the environment (Geyer et al., 2017). Oceans are particular dumping grounds for these products as plastic waste accounts for 80% of all marine litter (IUCN, 2021). Incinerating plastic waste is especially harmful because of the toxins released into the atmosphere. The problem of plastic waste is a persistent problem given that plastic bottles, for example, can take 450 years to decompose (Abdul-Rahman and Wright, 2014).

High-income countries, such as Canada, generate more plastic waste per person than on average (The World Bank, 2021) and should, therefore, have a greater responsibility for the management of plastic waste. Canadians make up less than 1% of the global population, but use 1.4% of all manufactured plastics (Young, 2021). Recycled plastics are often shipped abroad, predominantly to South Asian countries. This is for two main reasons: insufficient processing facilities in Canada and the ability to profit from waste sold to nations in need of resources. In 2015 approximately 3% of Canada's plastic waste, or 100,618 tons, was sold to China (de Montrouge, 2019). In 2018, China restricted its import of waste materials and Canada shifted its plastic waste exports to the United States, sending 101,131 tons to

[1] 2304 - 2240 Lakeshore Blvd W.Toronto, Ontario M8V 0B1.
[2] McMaster University, 1280 Main St. W., Hamilton On Canada.
Email: krantz@mcmaster.ca
* Corresponding author: leblanc.ryder@gmail.com

recycling brokers in the US (Banks, 2020). Although Canada has signed international agreements prohibiting the shipping of waste overseas, once Canadian waste ends up in the control of American exporters who can sell it to whomever they want, Canada is then no longer tracked by any regulations. As a consequence, plastic waste generated in Canada continues to be sent by the US to countries that do not have the facilities required to process it. This poses a threat to the health of people and the environment around the world. This policy loophole needs to be closed so that Canada can follow through on its international commitments (The Basel Convention) regarding the export of its plastic waste.

Current Regulations

The Basel Convention on the Control of Transboundary Movements is the only global legally binding agreement that specifically addresses plastic waste (BRS Secretariat, 2021). The Basel Convention has been an active instrument since 1992—originally negotiated under the United Nations Environment Programme in 1988—to protect human health and the environment by regulating the transboundary movement of hazardous substances. In 2019, it was amended to include plastic waste, effectively naming plastics as a hazardous material. The amendments identify plastics that are considered hazardous and expect countries to treat the transboundary movement of these substances, as they would with any other hazardous waste that has been covered in the convention. This means that parties to the convention have a legal obligation to ensure that their plastic wastes do not pose a threat to human health or the environment, including at their place of disposal. Specifically, countries are required to check with and ensure that the transport and disposal of plastic waste in importing countries is done in an environmentally sound way. Currently, 189 parties have signed the Basel Convention, including Canada and the US. Notably, however, the US never ratified the convention and is therefore not legally bound by it.

In addition to the Basel Convention, Canada has over 10 federal acts, regulations, and agreements governing plastic waste. These include the Microbeads in Toiletries Regulations (2017), the Ocean Plastics Charter (2018), and the Canada-wide Action Plan on Zero Plastic Waste (2018). Additionally, in May 2021, the Canadian Environmental Protection Act (1999) added "plastic manufactured items" to its Toxic Substances List (Canada, 2022). This action empowers the Government of Canada to make regulations that support its plastic waste goals.

Risks Associated with Plastic Waste Exports

Plastic waste and its by-products pose social, environmental, economic, and health risks. Social risks are tied to how the environment becomes unusable as it is polluted by plastic. Beaches and bodies of water that are traditionally used for recreation become less appealing as they fill with plastic waste. This results in people spending less time in such recreational areas which limits their ability to meet their social needs (Beaumont et al., 2019). Though illegal in Canada, plastic is incinerated in open fields in countries without proper facilities for waste disposal, which degrades the aesthetic value of the environment. This raw burning of plastic waste reduces air quality and deters individuals from spending time outside, further limiting their areas for social activity.

Plastic and potential contaminants pose significant environmental threats. Plastic, just by being present in the environment, affects carbon and nutrient cycles, disturbs habitat availability, entangles and suffocates wildlife, and puts additional stress on the populations of endangered or otherwise vulnerable species (MacLeod et al., 2021). Plastic is often used to store or transport other hazardous substances, and so contaminated plastic waste introduces additional threats to the environment.

Economically, plastics improve efficiency and benefit the food and transportation sectors, but these advantages come at a significant cost. Bodies of water with plastic waste impact fish populations and reduce the production of fisheries. Plastic waste also reduces the tourism and recreation appeal of the environment, reducing further revenues in these sectors as well. Reduced air quality slows overall economic output by causing premature deaths and ailments in the population. When these effects are taken together, it is estimated that the average cost of plastic waste globally is US $75 billion each year (UNEP, 2014).

Plastic waste is a threat to human health as it can either be ingested directly or its harmful components can find their way into the human body through other means. It is estimated that the average person consumes 5 grams of plastic each week, usually in the form of microplastics (MPs) in drinking water (WWF, 2019). Plastic waste that is poorly incinerated introduces new health risks through the toxins that are released during the process. These include Dioxins, Furans, Mercury, and Polychlorinated Biphenyls, which reduce air quality and expose humans to residues on crops and in waterways (Verma et al., 2016). Dioxins alone cause cancer and damage the neurological, reproductive, and respiratory systems (Verma et al., 2016).

Current Policy Obstacles/Challenges

Data shows that "it is the management of plastic waste that determines the risk of plastic entering the ocean" (Ritchie, 2018); also, medium- and low-income counties are more likely to mismanage waste because they lack the disposal infrastructure of wealthier countries. The 2019 plastic amendments to the Basel Convention were added to combat this very problem and as noted above, while the US also signed the Convention, it never ratified it and is, therefore, not bound by its regulations.

This asymmetry between Canada and the U.S. was addressed in October 2020 in the form of a bilateral agreement on plastic waste. Under the Basel Convention, parties are allowed to ship waste to non-signatory nations as long as they have a bilateral agreement. Though the Convention requires that any agreement meet or exceed the standards of the Convention itself, this particular agreement is not legally binding (Dyer, 2020). Therefore, once Canadian waste is shipped to American importers, there are no reliable mechanisms for tracking or ensuring the waste is disposed of properly or that it is even disposed of in the U.S.

Policy Evaluation

The US has recently adopted a national recycling strategy, and the Environmental Protection Agency has said it will explore the ratification of the Basel Convention (Calma, 2021). This suggests that the bilateral agreement may not be needed in the future but unfortunately, it is unlikely that the U.S. will ratify the convention or any legally binding agreement with Canada soon. The U.S. is increasing its shipments of plastic waste to developing nations (Steiner, 2021). This is unsurprising since a similar trend has been seen in the export of electronic waste from the U.S. (Lecher, 2019), which is another controlled substance under the Basel Convention. The fact of the matter is that waste exports in the U.S. are held to fewer standards than required under the Basel Convention and as long as the U.S. does not ratify the treaty, brokers will continue to do whatever they like with both domestic and imported Canadian waste. The statistics show that they prefer exporting to developing nations, which lack the capacity for proper disposal. Therefore, the 2020 bilateral agreement on plastic waste between Canada and the U.S. is insufficient for reaching Canada's zero plastic waste goal and contradicts the commitments enshrined in the Basel Convention.

Canada's other regulations on plastic waste are also insufficient for reaching its zero plastic waste goal. The Microbeads in Toiletries Regulations (2017), though useful for reducing Canada's microbead (plastic particles equal to or less than 5 mm in size) waste, only applies to toiletry products and does not address regular plastics. The Ocean Plastics Charter (2018) is insufficient because it is unenforceable and lacks the support of the U.S. Though it aims to bring together leading governments and organizations to commit to taking action to make plastics more sustainable and resource-efficient, the US is not a signatory. While the Charter encourages an increase in domestic capacity to manage plastics, which would lessen Canadian exports to the U.S., it does not state an end to plastic exports, which leaves this issue unaddressed. The Canada-wide Action Plan on Zero Plastic Waste (2018), a more structured and comprehensive policy on plastic waste, does not address plastic exports either. It does commit to ambitious targets of 100% circular (reusable, recyclable, or recoverable) plastics by 2030 and recovery of 100% of all plastics by 2040 but without any commitments on tracking exports, this may not address the problem. Producing and recovering more recyclable plastic will not help the plastic waste issue if all this waste is exported to the U.S. and then on to countries that cannot properly process it.

Conclusion

Canada has made promises on plastic waste under the Basel Convention and has over 10 domestic acts, regulations, and agreements. These commitments stand to be undermined by an unenforceable bilateral agreement with the US, which has not made the same commitments. Unfortunately, the U.S. does not seem prepared to take serious action on plastic waste, and therefore it is unreasonable to expect them to ratify the Basel Convention or any bilateral agreement that involves legal accountability. Therefore, the only way for Canada to act in line with its commitments is for it to revoke the bilateral agreement with the U.S. This action will create many questions about what is to be done with Canada's plastic waste, but these are questions that need to be answered and will likely not be considered so long as this bilateral loophole exists.

References

Abdul-Rahman, F. and Wright, S.E. 2014. Reduce, Reuse, Recycle: Alternatives for Waste Management. NM State University. https://aces.nmsu.edu/pubs/_g/G314.pdf

Banks, K. (2020, March 3). Where Canada sends its garbage. 10,000 Changes. https://10000changes.ca/en/news/where-canada-sends-its-garbage/

Beaumont, N.J., Aanesen, M., Austen, M.C., Börger, T., Clark, J.R., Cole, M., Hooper, T., Lindeque, P.K., Pascoe, C. and Wyles, K.J. 2019. Global ecological, social and economic impacts of marine plastic. *Marine Pollution Bulletin* 142: 189–195. https://doi.org/10.1016/j.marpolbul.2019.03.022

BRS Secretariat. (2021, January 1). BC Plastic Waste Amendments. Http://Www.Brsmeas.Org/. http://www.brsmeas.org/Implementation/MediaResources/PressReleases/BCPlasticWasteAmendments/tabid/8728/language/en-US/Default.aspx

Calma, J. (2021, November 15). The US finally adopts a national recycling strategy. The Verge. https://www.theverge.com/2021/11/15/22783450/recycling-united-states-epa-plastic-pollution-waste

Canada. (2022, September 6). Toxic Substances List: Schedule 1. Canada.ca. https://www.canada.ca/en/environment-climate-change/services/canadian-environmental-protection-act-registry/substances-list/toxic/schedule-1.html

de Montrouge, P.D. (2019, January 10). Media Briefing: Canada's Plastic Waste Export Trends Following China's Import Ban. Greenpeace Canada. https://www.greenpeace.org/canada/en/qa/6971/media-briefing-canadas-plastic-waste-export-trends-following-chinas-import-ban/

Dyer, E. (2020, December 3). Government quietly made "back door" agreement with U.S. that could undermine treaty on plastic waste. CBC. https://www.cbc.ca/news/politics/canada-us-plastic-waste-basel-1.5827340

Geyer, R., Jambeck, J.R. and Law, K.L. 2017. Production, use, and fate of all plastics ever made. *Science Advances* 3(7). https://doi.org/10.1126/sciadv.1700782

IUCN. (2021, November 17). Marine plastic pollution. https://www.iucn.org/resources/issues-briefs/marine-plastic-pollution

Lecher, C. (2019, December 4). The dark side of electronic waste recycling. The Verge. https://www.theverge.com/2019/12/4/20992240/e-waste-recycling-electronic-basel-convention-crime-total-reclaim-fraud

MacLeod, M., Arp, H.P.H., Tekman, M.B. and Jahnke, A. (2021, July 2). The global threat from plastic pollution. Pubmed. https://pubmed.ncbi.nlm.nih.gov/34210878/

Ritchie, H. (2018, September 1). Plastic Pollution. Our World in Data. https://ourworldindata.org/plastic-pollution#mismanaged-plastic-waste

Steiner, M. (2021, February 5). US-CANADA PLASTIC TRADE DEAL SUBVERTS BASEL CONVENTION. Basel Action Network. https://www.ban.org/news/2020/12/22/us-canada-plastic-trade-deal-subverts-basel-convention12

The World Bank. (2021). Trends in Solid Waste Management. Worldbank.Org. https://datatopics.worldbank.org/what-a-waste/trends_in_solid_waste_management.html

UNEP. (2014, June). Plastic Waste Causes Financial Damage of US$13 Billion to Marine Ecosystems Each Year as Concern Grows over Microplastics. UN Environment. https://www.unep.org/news-and-stories/press-release/plastic-waste-causes-financial-damage-us13-billion-marine-ecosystems

Verma, R., Vinoda, K., Papireddy, M. and Gowda, A. 2016. Toxic Pollutants from Plastic Waste—A Review. *Procedia Environmental Sciences* 35: 701–708. https://doi.org/10.1016/j.proenv.2016.07.069

WWF. (2019, June 12). Revealed: plastic ingestion by people could be equating to a credit ca. https://wwf.panda.org/wwf_news/press_releases/?348337/Revealed-plastic-ingestion-by-people-could-be-equating-to-a-credit-card-a-week

Young, R. (2021, November 26). Canada's plastic problem: Sorting fact from fiction. Oceana Canada. https://oceana.ca/en/blog/canadas-plastic-problem-sorting-fact-fiction/

Chapter 10

Nature-Based Solutions to Address the Plastics Problem

Biomimicry

Negin Ficzkowski and *Gail Krantzberg**

INTRODUCTION

The advent of plastic addressed many immediate needs for human society, incrementally creating a culture wherein eliminating it from the market is difficult to envision (Fig. 1). Consequently, many efforts aimed at combating its adverse effects are focused on replacing its feedstock more so than rethinking the underlying design. The persistent residues of petroleum as macro-, meso-, or micro-pollutants, the resulting degradation of the entire natural ecosystem, the limited non-renewable reserves of fossil fuels, and the resource-heavy extraction processes have encouraged a shift toward using renewable biomass sources, such as corn, cassava, sugar beet, sugar cane, or cellulose that offer potentials for reducing the overall environmental impacts of the produced waste (Tolinski, 2012). The increasing appetite for solutions in this category has created a market that is nearly saturated with what is commonly known as "bioplastics" which, contrary to popular belief, does not inherently address the emissions or pollution issues attributed to conventional fossil-based plastic.

The term bioplastic has not been regulated, and it is currently used to refer to a plastic material that contains some undefined percentage of biological-based feedstock, is biodegradable, or features both properties (European Bioplastics, 2022). This vague labelling enables the notion of green washing and poses a public policy challenge that not only encourages consumers to buy products that do not offer the promised environmental benefit (Wright, 2020) but also misleads them toward ineffective disposal practices. From a design standpoint, the pattern continues to follow the linear model of take-make-use-dispose, resulting in burden

McMaster University, 1280 Main St. W., Hamilton ON L8S 4K1.
* Corresponding author: krantz@mcmaster.ca

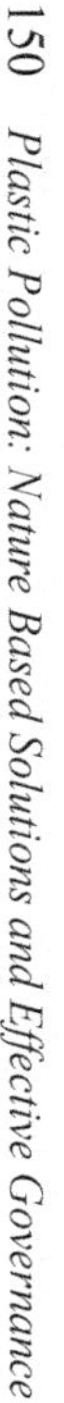

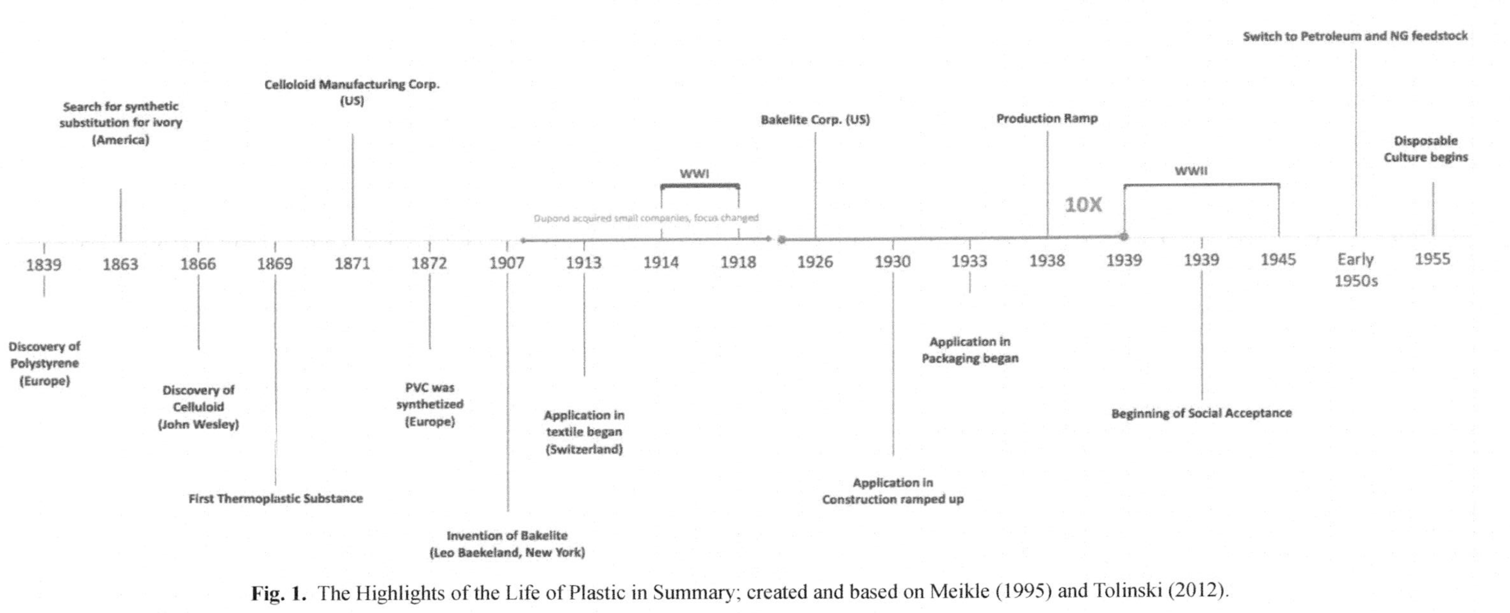

Fig. 1. The Highlights of the Life of Plastic in Summary; created and based on Meikle (1995) and Tolinski (2012).

shifting throughout the lifecycle of these products (Andrews, 2015; McDonough and Braungart, 2002). In the case of the plastic crisis, any effective solution should simultaneously consider the environmental and climate costs resulting from systemic complexities and failures of the plastic industry over the past century. This chapter investigates the opportunities this shift provides for Canada, the existing barriers, and the changes required within the Canadian landscape to allow for such innovative ventures to become mainstream.

Experts suggest that in addition to efforts in replacing the conventional raw materials, marketing incentives and promoting products that are multiuse are necessary for addressing the inefficacy of global waste management practices including the plastic's (Fig. 2). Reducing the consumption of plastic involves a radical and transformative shift that includes a paradigm shift for the government, businesses, and society from immoderate consumerism and linear consumption patterns to a sustainable circular and regenerative lifestyle (Learn Biomimicry, 2021). Incremental interventions are necessary for slowing down the crisis; a hasty underdeveloped shift away from conventional plastic and toward the next best alternative without a shift in the mental model can cause more harm than good. For instance, replacing the famously debated plastic straws with metal straws at similar consumption rates introduces higher carbon emissions while remediating less than 1% of the ocean plastic waste crisis. Similarly, replacing liquid-carrying plastic

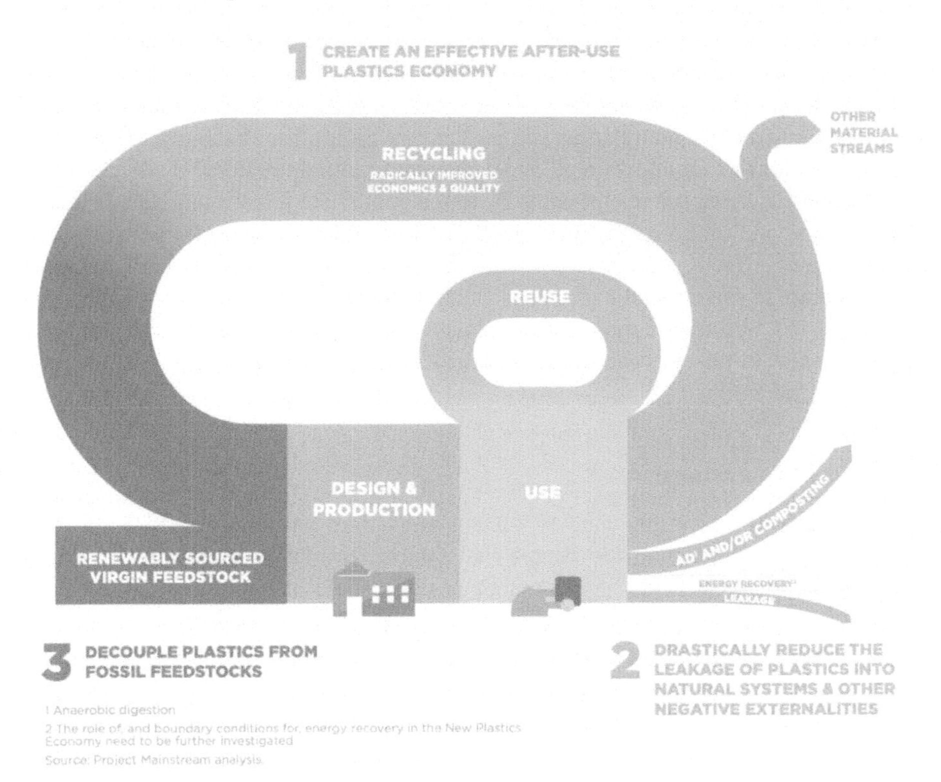

Fig. 2. The Ambitions of the New Plastic Economy (Source: Ellen Macarthur Foundation, 2016).

containers with resource-heavy materials such as glass or aluminum without having an effective reuse business model or recycling infrastructure in place can result in higher global warming potential over the lifecycle of the products.

Since 1950, a mere 9% of the plastic ever produced has been recycled back into the economy (Wuennenberg and Tan, 2019). This number is shown to have reduced even further since the start of the global COVID-19 pandemic by more than 20% in Europe, 50% in Asia, and about 60% in some parts of the US, while the pandemic-associated plastic waste was estimated to be over eight million tons (Peng et al., 2021). Many efforts are underway globally for scaling up the rate of plastic recycling, hoping to reduce the amount of plastic waste entering the environment; but since almost none of the conventional products are designed for circularity, the recovery rates and the externalities of these efforts (such as energy use, water footprint, and emissions from transportation) vary significantly (Liboiron, 2012).

It is estimated that over 80% of all product-related environmental impacts are determined during the design phase of a product (IISD, 2011), therefore developing policies that target responsible design of the products and services can provide a pathway to alternative concepts and strategies for sustainable development in consideration of the long-term impact on climate, society, and the planet. As an emerging field, responsible design theory challenges the traditional definition and practice of design to embed a planet-centric, pluriversal, and transdisciplinary approach in proposing solutions and interventions (Escobar, 2018). Many emerging design attitudes began recognizing this socio-technical complexity and aim to embrace planetary diversification into the development of science and technology, which needs radically different ways of observing the world and understanding how all entities are interrelating and influencing each other (Iwabuchi, 2020). An alternative way to rethink, reframe, and respond to a multidimensional challenge, such as the plastic crisis, is made possible by looking into nature and natural ecosystems for inspiration. Nature-based Solutions (NBS) is an umbrella term designated to the applications of ecosystem services, natural capital, and lessons that can be learnt from nature to solve specific problems or challenges of human well-being (Potschin et al., 2016). Soft engineering approaches, green infrastructure, and biomimicry are all examples of this approach wherein the identification of the core problem is key (EC, 2015).

What is biomimicry?

Biomimicry is a tool that enables a move away from petroleum-based plastics if it is well integrated into the educational system, regulatory frameworks, design policies, and product management strategies. The field of biomimicry focuses on commercial mimicry of nature's ingenuity by recognizing the potential for emulating its strategies in the design and development of sustainable solutions, as long as these solutions cause no harm to nature and contribute to its conservation (Neves and Francke, 2012). Defining the right question is central to maintaining circularity through biomimicry or any other nature-based interventions (Potschin et al., 2016). The principles of biomimicry need to be applied at a systems level and for the design of the solutions in an integrated, long-term-oriented, and transformative way to enable the shift toward circular technological development (Escobar, 2018) and sustainable economic growth that supports further scientific understanding of the world (Andrews, 2015).

Historical Evolution of Biomimicry in Addressing the Linear Design Framework

The political climate within the oil-rich countries and the resulting rising oil prices in the 1970s became a driver for reinvesting in the research and development of bioplastics (British Plastics Federation, 2014). The majority of inventions in this era were focused on producing plastic polymers using microbes and bacteria. A notable discovery in this era was the principle of biodegradable plastic by Japanese scientists. Biodegradability has been done by adding chemicals or natural organisms into the plastics to assist in a quicker breakdown. While biodegradability introduces a promising solution for reducing the buildup of waste at landfills, it does not contribute to the reduction in MP sediments or bioaccumulation in the environment.

In 1996, Monsanto led a shift in the market by using plants instead of biological organisms in producing bioplastics (British Plastics Federation, 2014). Soon after, many plastic industry leaders joined forces to scale up plant-based plastic production. Certain plants such as seaweed immediately emerged as biomass with huge economic potential for many developing countries, such as Indonesia whose seaweed export was valued at around US $200 million in 2014 and was estimated to grow by 30% annually in plastic production (Barrett, 2018). Other plastic alternatives in the current market are made from the waste recovered from common agricultural processes globally, such as Avocado pits used as the main feedstock for Mexican-based BioFase for creating biopolymers (Biofase, n.d.) or the extracted fibres of pineapple leaves for the UK-based synthetic leather Piñatex (Ananas Anam, n.d.).

Pure plant-based plastics release carbon dioxide (or methane) and water when they break down (Tolinski, 2012). However, similar to the conventional plastic production processes, chemicals are typically added during the manufacturing process of conventional plant-based alternatives for increasing strength, preventing wrinkling, or conferring breath ability (European Bioplastics, n.d.). If added, toxins are released during degradation, impacting the environmental integrity of plant-based products. Furthermore, the addition of conventional adhesive for attaching plastic components or labelling them impact the rate of recovery after use. The products designed to be "compostable" within Canadian (and EU) standards have the capability to fully break down into a natural soil-like organic matter (Government of Canada, 2008). As such, by regulation, they cannot contain toxins that could leach into the soil (European Bioplastics, n.d.).

These challenges lead to biomimicry as a new direction for addressing the plastic pollution problem in the scientific literature. Emulating nature for design solutions is not new but an emerging discipline of ancient practices found within many native cultures (Learn Biomimicry, 2021). Biomimicry became widely known in the 1980s when George de Mestrel invented Velcro based on emulating the way seeds/burs stick to animal fur. In the 1990s many more isolated cases of bio-inspired products, processes, systems, and strategies emerged around the world (Learn Biomimicry,

2021). In 1997, science writer Janine Benyus provided an overarching perspective and framework for this field in her book on Biomimicry which became a catalyst in reintroducing the field in practice (Learn Biomimicry, 2021). Biomimicry life's principles focus on adapting to change within the larger context of earth's operating principles and the capacity of organisms to grow and develop. The development of advanced microscopes and analytical tools over the last decade has increased scientific understanding of the natural ecosystem and allowed biomimicry to tap into the knowledge base for translating nature's design principles into new concepts for solving emerging problems (Dorfman, 2020). Biomimicry case studies fundamentally emulate the natural organisms at form, process, or system levels that may fit the purpose (Learn Biomimicry, 2021). Spending time in nature provides learners with hands-on opportunities to experience natural design patterning, interactions among species, and processes that provide material regeneration (Cash, 2015). It is also repeatedly highlighted by academic experts that reflection and repeatability in design strategies are integral to a successful transfer of fundamental concepts into different applications and individual worldviews (Learn Biomimicry, 2021).

Practical Adaptation of Biomimicry in Addressing the Problem

Biomimicry systems thinking requires integrative policies that encourage the successful uptake of the theory and invention protection of resulting technologies. The examples below elaborate on the use of biomimicry life's principles to address the primary global challenges of plastic from immediate remediation to long-term integration:

a) Solutions for Cleaning Up Plastic Waste That Enters Waterways: Overwhelmed by the scale of the plastic problem, by 1986 many grassroots organizations began ocean plastic cleanup projects in order to collect and document the types of trash littering at the beaches and some coastal waterways (Mironenko and Mironenko, 2020). Many organisms in nature provide mechanisms that can be adopted for the effective cleanup of MP sediments. For example, Floating Coconet designed in the Netherlands aims to capture free-flowing plastics of all dimensions (micro-, meso-, or macro-) from rivers prior to entering the oceans through a mechanism inspired by filtration system in Manta Rays and basking sharks filter food from water (Ask Nature, 2016; 2019).
b) Managing Plastic After Its Initial Intended Use: Nature builds (and recycles) from the bottom up. This can translate into what we know as additive manufacturing or strategies for enabling a better rate of recycling for emerging plant-based solutions:
    - The Challenge of Adhesives: The physiology of Geckos enables them to climb a variety of surfaces as though they have adhered to them. Their toe pads are covered in millions of small hair-like projections that branch further into hundreds of nano-scale structures that end in tiny discs, giving them a high surface area. The tendons of the gecko's feet are stiff and stretch from the bone to the skin all the way to the underside of the feet, which allows for even distribution of forces when they stiffen their feet, hence sticking

to surfaces effortlessly. The Ford automotive company has introduced a mechanism inspired by this function that allows surface adhesion without using glue in automotive manufacturing. This allows for the complete disassembly of plastic parts and increased recycling rates. Other companies, such as Geckskin, have also adopted the system of gecko feet to create a strong adhesion device that allows for conforming objects to surfaces (such as labels on bottles) without leaving residue (Ask Nature, 2012; 2017; 2020).
- The Challenge of the Inner Lining of Plant-Based Containers: A company called Humble Bee in the US has developed a cellophane-like cell lining inspired by bee nesting lining processes and materials to improve the organic recovery of plant-based plastics (Humble Bee, 2018).
- The Challenge of Toxic Additives for Producing Colour: Having been inspired by plants and animals such as chameleon, various studies suggest changing the structure and thickness of material allows for eliminating colour additives in plastic production. However, none of these solutions have been transferred to the market yet due to increased cost and time of production (Ask Nature, 2020).

c) Using Less Material by Design: Some inventions use mechanisms found in nature to reduce the amount of plastic required in producing conventional items. There are deep patterns in nature that can inspire optimizing product structure with minimal materials. These patterns can be summarized in 12 categories focused on composition and the architecture of the materials, such as form following flow, functional gradients, cylindrical strength, and tensegrity (Benyus J., 2017). An example of this adaptation is the Portuguese-designed Vitalis Water Bottles which focuses on reducing the amount of material used per plant-based liquid bottle by providing lightweight and durability inspired by spiral patterns of pine tree trunks (Ask Nature, 2017; Biomimicry Institute, 2019).

d) Mindset Shift in Design: By questioning the fundamental principle of the industrial revolution and the unnecessary economic and environmental costs associated with conventional interior design practices, i2 carpet tiles mimic the random patterns of the forest floor by incorporating natural elements of the forest. The market shift created by this design enabled a significant reduction in waste generation, unplanned costs for enterprises, and embodied carbon associated with the "attic stock" of conventional carpets (Ask Nature, 2009). Many solutions have no technical component and are merely intended to influence consumer habits, such as supermarkets that are designed specifically to allow consumers to purchase products without any packaging (Learn Biomimicry, 2021).

e) Using Waste From One System as Raw Material for Another: Globally, 85% of the plastic waste in the earth's waterways consists of single-use and packaging items, highlighting the immediate need for applicable alternatives (Ellen Macarthur Foundation, 2016). Some solutions have been extremely simple and grassroots as these have focused on localized bioutilization for just-in-time packaging solutions, such as the traditional use of the green leaves of miracle fruit for packaging hot food in Nigeria and these leaves take less than six months after use to decompose back into nutrient-rich soil (UNDP, 2019). They can also involve systems-level biomimicry with limited processing of natural resources to mimic material cycling in nature and allow extended useful life of leaves for countries with limited immediate access, such as the German food packaging producer using pressed and dried leaves stitched into the form of simple food containers or the UK-based single-use tableware items and party supplies for hot and cold food made from naturally fallen Areca palm leaves in Southern India, which is gathered and cleansed with high-pressure water jets. Once dried, these leaves are shaped under heated moulds and ready for commercial use. There are no coatings, additives, or chemicals used, just the natural leaf. Similarly, Indian-based edible utensils are made from limited processing of sorghum flour commonly grown in South Asia, Africa, and Central America. Another example is the New York-based company, Ecovative which is using mycelium, the vegetative branching part of a fungus, and cellulosic agricultural byproducts, such as corn stalks, hemp hurds, and wood chips to make home compostable thermally insulating, and water-resistant packaging material, tiles, and planters (Ecocative, n.d.). The mycelium added to the cellulosic materials strongly bonds them together, replacing the need for moulded packaging.

More unconventional yet highly technical solutions are also on the rise to enable more cost-effective solutions on a global scale and transform natural "waste" into useful products that are fully biodegradable at the end of their life. To avoid the use of any nutritionally important materials and reduce the high cost of producing biobased alternatives like polyhydroxyalkanoates (PHAs), researchers at Columbia University are developing systems to intake wastewater (human waste) for producing biodegradable bioplastics. Similarly, research efforts from Japan and Stanford University in the US focus on material-process combinations to convert greenhouse gases from wastewater treatment plants or landfills into lactic acid or sugar for creating raw biomaterial for bioplastic production that biodegrades back into methane for natural digestion in marine microorganisms. Similar innovations use plant-driven sugars and fermentation processes to produce protein chains that mimic spider silk for less environmentally harmful textile production. A European study is investigating the potential for using animal waste material from slaughterhouses to synthesize plastic. The environmental and moral disputes regarding the meat industry aside, solutions like this are attempting to regard all waste as useful material and put them back into the flow. Such approaches would only constitute viable strategies for producing environmentally safe and efficient alternatives to conventional plastic if the processing steps do not require sophisticated lab procedures or financial resources and would not contribute relatively higher emissions and resource usage (water and energy) throughout their entire lifecycle.

Some inventions are attempting to replace the need for conventional packaging practices by producing a variety of new protective materials to be used for the purpose of transportation. For instance, a Japanese design company is producing a solution to provide enough cushioning function from agar in red marine algae for items to be transported (Pangburn, 2016). This solution has been developed based on a locally common dessert made by mixing agar and hot water that provides a cushion-like function after being frozen.

Discussion

Biomimicry systems' thinking recognizes that regardless of the scale of a problem, the big picture of the solution is rarely static and it focuses on creating patterns for leading change over time. The leverage points shown in any complex system intervention are the opportunities where a small shift can produce big changes (Davelaar, 2021). Constraints, feedback loops, and paradigm shifts are three levels of interventions (Meadows, 1999) that can encourage the shift from the status quo to a more nature-inspired mindset for resolving the plastic crisis using biomimicry thinking.

Canada along with France, Germany, Italy, the UK, and the European Union adopted the Ocean Plastics Charter on June 9, 2018, as an informal non-binding agreement (ECCC, 2018) to demonstrate its commitment to a more resource-efficient plastic economy (ECCC, 2020). Canada also has an integrated management approach to plastic products in the discussion that can benefit from the inclusion of time-bound targets for reducing conventional plastic production and from the increasing reuse/reusable models across sectors (Greenpeace, 2020).

The concept of the circular economy is currently being used in different jurisdictions as an incentive to better manage costs associated with end-of-life management. In addition to slowly progressing regulations and bans, market drives and financial incentives often attract increased participation for accelerating the adaptation of new methodologies and products (Yock et al., 2015). The application of biomimicry systems thinking is not yet widely recognized within certain professions (such as engineering design in an ecological context) and is slow compared to many other developed countries around the world (Cash, 2015). That said, some networks and private institutions, such as Biomimicry Alberta, Biomimicry Ontario, Biomimicry Commons, and Biomimicry Frontiers in Guelph, Ontario, are created to promote biomimicry education, mostly for integration of the principles in urban development. In addition to various academic research laboratories across the country, a few youth design challenges have embodied biomimicry applications in Alberta and British Columbia in various fields, such as water management and transportation.

In the US, biomimicry is recognized as a valuable design methodology technique for promoting a circular economy and minimizing the material impact on the environment (Cash, 2015). One of the most impactful shifts that has occurred in more advanced economies, such as the US or the UK, for addressing the plastic problem has been focused on business model developments that offer product services instead of products (Biomimicry Innovation Lab, 2021). Particularly, in Europe adapting biomimicry principles has progressed more due to the advances in the distributed economy as opposed to the centralized manufacturing systems in North America (Learn Biomimicry, 2021).

Conclusion

Nature is a model, measure, and mentor for regenerative and resilient development. The inspirations are everywhere and the technology is available. The key to applying these inspirations to tackling complex and multifaceted problems, such as that posed by plastic which is intertwined with every aspect of the economy, is incorporating system thinking tools and biomimicry life's principles rather than specific technologies or biological models.

References

Ananas Anam. (n.d.). *Piñatex*. Retrieved from Ananas-Anam: https://www.ananas-anam.com/about-us/

Andrews, D. 2015. The circular economy, design thinking and education for sustainability. *Local Economy* 30(3): 305–315.

Barrett, A. (2018, July 5). The History of Bioplastics. *Bioplastics News*.

Benyus, J. 2008. A good place to settle: Biomimicry, biophilia and the return of nature's inspiration to architecture. pp. 27–42. *In*: S. Kellert, J. Heerwagen and M. Mador (eds.). *Biophilic Design, The Theory, Science, and Practice of Bringing Buildings to Life*. Hoboken, NJ: John Wiley & Sons, Inc.

Benyus, J. (2017, November 17). Lightweighting Inspired by Nature (Seminar). Biomimicry 3.8.

Biofase. (n.d.). *Technology* . Retrieved from Biofase: https://biofase.com.mx/technology

Bioplastics Magazine. (2017, November 13). *Scientists are trying to synthesize environmentally-friendly plastic alternatives, using cyanobacteria*. Retrieved from https://www.bioplasticsmagazine.com/en/news/meldungen/20171113_cyanobacteria.php

British Plastics Federation. 2014. *A History of Plastics*. Retrieved from bpf.co.uk

Canada Business Registry 2022.

Cash, K. 2015. Beyond LEED®: Constructing a Bridge to Biomimicry for Canadian Interior Design Educators. Winnipeg, Manitoba, Canada: Thesis submitted to the Faculty of Graduate Studies, The University of Manitoba.

Chambers, M. 2011. Enhanced searchable database for exploration—Application of biomimicry in interior design. *Graduate Thesis and Dissertations*. Iowa State University, Paper 11958.

Cocker, J., Pariseau, J. and Larnder-Besner, M. (2021, March 19). State of regulation of plastics in Canada: The basics. *BLG Law*.

Da Silva, S. 2021. *Towards Osaka Blue Ocean Vision: National Action Plan*. Ministry of the Environment, Japan.

Dale, A. and Newman, L. 2005. Sustainable development, education and literacy. *International Journal of Sustainability in Higher Education* 6(4): 351–362. doi:10.1018/14676370510623847.

Damstra, J.R. (2020, February 5). *Supreme Court of Canada Silent on Environmental Regulation Cases*. Retrieved from https://www.lerners.ca/lernx/climate-cases/

Davelaar, D. 2021. Transformation for sustainability: a deep leverage points approach. *Sustainability Science* 16: 727–747 .

Dorfman, M. (2020, April 20). Biology Inspires a Plastics (R)evolution. Synapse. Biomimicry 3.8.

Ecocative. (n.d.). Retrieved from https://ecovative.com/why-mycelium

Ellen Macarthur Foundation. 2016. *The New Plastics Economy: Rethinking the Future of Plastics*.

Escobar, A. 2018. *Designs for the Pluriverse*. North Carolina: Duke University Press.

ECCC. 2018. *Ocean Plastics Charter*. Retrieved from Government of Canada: https://www.canada.ca/en/environment-climate-change/services/managing-reducing-waste/international-commitments/ocean-plastics-charter.html

ECCC. 2020, October 7. *Canada one-step closer to zero plastic waste by 2030*. Retrieved from Government of Canada: https://www.canada.ca/en/environment-climate-change/news/2020/10/canada-one-step-closer-to-zero-plastic-waste-by-2030.html

Environment and Climate Change Canada. (2020, October 10). A proposed integrated management approach to plastic products to prevent waste and pollution: Discussion Paper. *Order Adding a Toxic Substance to Schedule 1 to the Canadian Environmental Protection Act, 1999:* 154(41).

Eunomia. 2020. *What is a Plastic? A summary report exploring the potential for certain materials to be exempted from the Single-Use Plastics Directive.* Eunomia Research & Consulting.

Fawcett-Atkinson, M. (2021, March). *Canada is drowning in plastic waste and recycling won't save us.* Retrieved from Canada's National Observer investigation.

Public Policy Forum. 2020. *Nature-based Solutions: Some of the answers to climate change come naturally.*

Fry, T. 2009. *Design futuring.* New York, NY: Berg Publishers.

Government of Canada. (2019, August 22). *Heritage Institutions Across British Columbia Receive Support from the Government of Canada.* Retrieved from Canadian Heritage: https://www.canada.ca/en/canadian-heritage/news/2019/08/backgrounder--heritage-institutions-across-british-columbia-receive-support-from-the-government-of-canada.html

Government of Canada. 2021. *Competition results for the Discovery Accelerator Supplements program.* Retrieved from Natural Sciences and Engineering Research Council of Canada: https://www.nserc-crsng.gc.ca/NSERC-CRSNG/FundingDecisions-DecisionsFinancement/DiscoveryAcceleratorSupplements-SupplementsAccelerationDecouverte/index_eng.asp?Year=2021

Government of Canada. (2021, February 3). *Bill C-204, An Act to amend the Canadian Environmental Protection Act, 1999* .

Government of Canada. (June 2008). Environmental Claims: A Guide for Industry and Advertisers (Archived)

Government of Canada. (2022, January 20). Environmental claims and greenwashing.

Grasso, M. (2019). Oily politics: A critical assessment of the oil and gas industry's contribution to climate change. *Energy Research & Social Science* (50): 106–115 https://doi.org/10.1016/j.erss.2018.11.017.

Greenpeace. (2020, December). *Plastic Recycling: That's not a thing.* Retrieved from https://www.greenpeace.org/static/planet4-canada-stateless/1d30117a-greenpeacereport_plasticrecyclingthatsnotathing.pdf

Hamilton, L.A., Feit, S., Muffett, C., Kelso, M., Rubright , S. M., Bernhard, C., . . . Labbé-Bellas, R. 2019. *Plastic & Climate: The Hidden Costs of a Plastic Planet.* Center for International Environmental Law (CIEL).

Humble Bee. 2018. *Humble Bee Bio.* Retrieved from https://www.humblebee.co.nz

IEA. 2018. *The Future of Petrochemicals.* OECD/IEA.

Innovations & Partnership Office. 2021. Inventor's Guide to Technology Transfer. 2.

IISD. 2020. *Unpacking Canada's Fossil Fuel Subsidies.* Retrieved from International Institute for Sustainable Development: https://www.iisd.org/articles/unpacking-canadas-fossil-fuel-subsidies-faq

Iwabuchi, M. (2020, July 3). *Emerging Design Attitudes: Speculative, Transitional, and Pluriversal Design: Go beyond Cartesian Belief System in this Century.* Retrieved from UX Planet: https://uxplanet.org/design-attitudes-for-this-century-speculative-transitional-and-pluriversal-design-fb55c9d401e6

Learn Biomimicry. 2021. Biomimicry Master Class Course Material. Biomimicry Switzerland.

Liboiron, M. (2020, November 19). *Research on the relationship between EPR and shoreline plastics.* Retrieved from CLEAR: https://civiclaboratory.nl/2020/11/19/research-on-the-relationship-between-epr-and-shoreline-plastics/

Liboiron, M. (2012, July 25). *Designing a reuse symbol and the challenge of recycling's legacy.* Retrieved from Discard Studies: https://discardstudies.com/2012/07/25/designing-a-reuse-symbol-and-the-challenge-of-recyclings-legacy/

McDonough, W. and Braungart, M. 2002. Design for the triple top line: new tools for sustainable commerce. *Corporate Environmental Strategy* 9(3): 251–258.

Meikle, J.L. 1992. Into the Fourth Kingdom: Representations of Plastic Materials, 1920-1950. *Journal of Design History* 1992, 5(3): 173–182.

Meikle, J.L. 1995. *American Plastic: A Cultural History.* Rutgers University Press.

Meadows, D.H. 1999. *Leverage Points: Places to Intervene in a System.* Sustainability Institute.

Mironenko, O. and Mironenko, E. (2020, September 16). Education Against Plastic Pollution: Current Approaches and Best Practices. pp. 67–93. *In*: F. Stock, G. Reifferscheid, N. Brennholt, E. Kostianaia (eds.). *Plastics in the Aquatic Environment - Part II. The Handbook of Environmental Chemistry* (Vol. 112). Springer, Cham. https://doi.org/10.1007/698_2020_486.

Natural Sciences and Engineering Research Council of Canada. (n.d.). *Plastics science for a cleaner future*. Retrieved from https://www.nserc-crsng.gc.ca/Professors-Professeurs/RPP-PP/Plastics-Plastiques_eng.asp

Pangburn. (2016, September 03). New Seaweed-Based Material Could Replace Plastic Packaging.

Potschin, M., Kretsch, C., Haines-Young, R., Furman, E., Berry, P. and Baró, F. 2016. Nature-Based Solutions. *In*: M. Potschin and K. Jax (eds.). *OpenNESS Ecosystem Services Reference Book.* Available via: www.openness-project.eu/library/reference-book: EC FP7 Grant Agreement no. 308428.

Roland, E.C. 2009. 8 Forms of Capital. *Permaculture Magazine #68*, pp. 58–61.

Rochman, C. (2020, September). The story of plastic pollution: from distant ocean gyres to the global policy stage. *The oceanography Society* 33(3): 60–70.

Ryan, P.G. 2015. A Brief History of Marine Litter Research. pp. 1–27. *In*: M. Bergmann, M. Klages and L. Gutow, *Marine Anthropogenic Litter* (. https://doi.org/10.1007/978-3-319-16510-3_1)). Springer International Publishing AG Switzerland.

Strasser, B. and Schlich, T. 2020. The art of medicine: A history of the medical mask and the rise of throwaway culture. *The Lancet*, https://doi.org/10.1016/S0140-6736(20)31207-1 1.

The European Economic and Social Committee. 2018. *A European Strategy for Plastics in a Circular Economy COM/2018/028*. Brussels: European Commission.

Thompson et al., R.S. 2009. Our plastic age. Philosophical Transactions of the Royal Society B. 364: 1973–1976 doi:10.1098/rstb.2009.0054.

Tolinski, M. 2012. *Plastics and Sustainability: Towards a Peaceful Coexistence between Bio-based and Fossil Fuel-based Plastics.* Hoboken, New Jersey; Salem, Massachusetts.

Wright, G. (2020, Janauary 29). 83% of shoppers mislead by green & sustainable advertising. *Retail Gazette*.

Wuennenberg, L. and Tan, C.M. 2019. *Plastic Waste in Canada: A daunting economic and environmental threat or an opportunity for sustainable public procurement?* IISD.

van Sebille, E., Wilcox, C., Lebreton, L., Maximenko, N., Hardesty, B.D., Franeker, J.A. and Law, K.L. 2015. A global inventory of small floating plastic debris. *Environmental Research Letters* 10: doi:10.1088/1748-9326/10/12/124006.

Walker, T.R., McGuinty, E., Charlebois, S. et al. 2021. Single-use plastic packaging in the Canadian food industry: consumer behavior and perceptions. *Humanit. Soc. Sci. Commun* 8: 80. https://doi.org/10.1057/s41599-021-00747-4

Webb, J. 2012. *Climate Change and Society: The Chimera of Behaviour Change Technologies.* UK: sagepub DOI: 10.1177/0038038511419196.

Yock, P., Zenios, S., Makower, J., Brinton, T., Kumar, U., J, W. and Kurihara, C. 2015. *Biodesign.* Cambridge University Press.

Chapter 11

Microplastic Research in India: Current Status and Future Perspectives

Arunbabu V.[1] and *E.V. Ramasamy*[2,*]

INTRODUCTION

Plastics are undoubtedly one of the most important inventions of the twentieth century. Because of their unique properties such as lightweight, high strength, durability, ease of molding into any shape, low cost of production, etc., plastics have found application in almost all aspects of human life, including health care. Consequently, the global production of plastics has increased from 2 million tons in 1950 to 380 million tons in 2015 and is projected to increase in the coming decades (Geyer et al., 2017). However, the proliferation of single-use plastics (SUPs) has put the waste management system under pressure. It is estimated that approximately 6,300 million tons of plastic waste has been generated globally between 1950 and 2015, and 79% of this has been accumulated in landfills and the natural environment (Geyer et al., 2017). It is further estimated that if current production and waste management trends continue in business as usual (BAU) mode, approximately 12,000 million tons of plastic waste will be in the natural environment and landfills by 2050. Another estimate indicated that 4 million to 12 million tons of plastic waste entered the marine environment in 2010 alone (Jambeck et al., 2015). The majority of the plastic debris in the oceans arises from land-based sources. Lebreton et al. (2017) estimated that rivers transport 1 million to 2.5 million tons of plastic to the seas every year.

Microplastics (MPs), which are defined as plastic particles < 5 mm in size, are emerging as a contaminant of global concern. They are ubiquitous and pervasive in the environment, and their presence has been detected in lakes (Sruthy and Ramasamy, 2017), rivers (Mani et al., 2015), oceans (Andrady, 2011), terrestrial ecosystems (Beriot et al., 2021; de Souza Machado et al., 2018), and atmosphere (Chen et al. 2020). MPs

[1] School of Environmental Studies, Thunchath Ezhuthachan Malayalam University, Tirur, Malappuram, Kerala 676 502, India.
Email: arun@temu.ac.in

[2] School of Environmental Sciences, Mahatma Gandhi University, Priyadarsini Hills P.O., Kottayam, Kerala 686 560, India.

* Corresponding author: evramasamy@mgu.ac.in

are even reported from remote areas, such as the Arctic (González-Pleiter et al. 2020; Ramasamy et al. 2021) and Antarctic regions (Waller et al., 2017). MPs originate from both primary and secondary sources. The primary sources include microbeads used in personal care products, fibers used in textiles, and pellets used as raw materials in manufacturing plastic products. The secondary MPs result from the fragmentation of larger plastics in the environment due to photo-oxidation, thermal degradation, mechanical action of waves and wind, and biological actions. The major sources of plastics and MPs in the environment come from unscientific waste disposal, discharge of treated and untreated domestic and industrial wastewater, stormwater runoff, road and tire wear particles, atmospheric deposition, shipping, etc.

Many recent studies reported the ingestion of MPs by aquatic and terrestrial organisms, including humans (Daniel et al., 2020a; Kutralam-Muniasamy et al., 2020; Schwabl et al., 2019). However, the consequences of MP ingestion by aquatic and terrestrial organisms are not well understood. The ingested MPs can significantly impact organisms through food intake and digestive system impairment. The additives in plastics can also leach into the digestive system of organisms. Apart from this, the MPs have been reported to adsorb many chemicals, including persistent organic pollutants and heavy metals from the surrounding environment (Dong et al., 2020). Therefore, MPs can act as vectors of toxic contaminants in organisms. All these factors signify the requirement for stringent control measures and policies to reduce plastic and MP pollution.

With more than 1.2 billion population, India is the second-most populous country globally. It is also one of the major consumers of plastic items. However, due to inadequate infrastructure and poor waste management practices, huge quantities of plastic waste are littered in the environment. It is estimated that in 2018, approximately 3.3 million tons of plastic waste were generated in India (CPCB, 2019). The lack of public awareness and inadequate infrastructure to collect and treat municipal solid waste leads to the littering of plastic waste in every nook and corner of the country. Fragmentation of these plastics leads to the formation of MPs. The discharge of untreated municipal wastewater also contributes significantly to MP pollution in the environment.

Estimating MP contamination in various environmental matrices and examining the sources of such MPs is necessary to develop suitable policies to mitigate plastic waste and evaluate the effectiveness of such policies. In this context, selected studies from India reporting MPs in various environmental matrices, biota, and food items are reviewed in the present study. The concentration of MPs in various ecosystems and biota is highlighted, and the impact on human health is discussed. The present policies on plastic waste management are briefly discussed. The major challenges facing MP research in India are highlighted, and the priority areas for future research are proposed as well.

MPs in Rivers, Lakes, and Wetlands

The contamination of rivers, lakes, and wetlands by MPs is a serious concern as the local people depend on these ecosystems for water, food, and livelihood. These aquatic ecosystems also reflect the level of contamination in the terrestrial environment. The rivers act as a major transport route for MPs into the marine environment. Napper et al. (2021) estimated that 1–3 billion pieces MPs are discharged into the Bay of Bengal through the Ganges every day. As these aquatic ecosystems play a significant role in groundwater recharge, the presence of MPs in the sediments of these ecosystems also has implications for groundwater contamination. Despite these facts, the contamination of freshwater ecosystems with MPs has received less attention globally than marine ecosystems. Despite having a large number of rivers, lakes, and wetlands in India, very few studies were conducted on MP pollution in these ecosystems.

The first Indian lake to be investigated for MP contamination was Lake Vembanad in Kerala (Sruthy and Ramasamy, 2017). Later, Lake Red Hills (Gopinath et al., 2020) and Lake Veeranam (Bharath et al., 2021) in Tamil Nadu and Lake Renuka (Ajay et al., 2021) in Himachal Pradesh were investigated for MP contamination (Table 1). The surface water and sediment samples were analyzed from these lakes, except for Vembanad, where only the sediment was analyzed. From the reported literature, it is observed that MPs are present in all the lake samples. Hence, it could be inferred that MPs are pervasive pollutants in the lakes of India. The common morphological classifications of MPs reported in these studies are fibers, films, fragments, and pellets. Fourier-transform infrared spectroscopy (FTIR) and Raman Spectroscopy were used to identify the polymers. The polymers such as polyethylene (PE), polypropylene (PP), and polystyrene (PS) are reported in all the studies, thus indicating the wide use of these polymers across the country. In addition to these polymers, polyvinyl chloride (PVC) and nylon were reported from Lake Veeranam (Bharath et al., 2021). Most of the studies targeted MPs in the size range of 0.3 mm to 5 mm. The abundance of MPs in water and sediment samples of Lake Renuka varied from 2–64 particles per liter and 15–632 particles per kg respectively (Ajay et al., 2021). The abundance of MPs in most studies falls within the range reported from Lake Renuka.

MPs of different morphologies, such as fragments, fibers, foams, films, pellets, and microbeads are reported from Indian rivers (Table 1). The different polymers reported are PE, PP, PS, PA, PET, PVC, polyester, nylon, acrylic, etc. Most studies reported MPs in the size range of 0.3 mm to 5 mm. However, there are disparities in selecting the size range of MPs among different studies. Few studies reported MPs as small as 20 μm. The MPs in the river sediments range from 17 to 3,485 particles per kg (Table 1).

The presence of low-density polymers, such as PE and PP, in the lake and river sediments indicates biofouling and adsorption of contaminants over the surface of MPs, which increases their density leading to subsequent sedimentation. Most studies reported land-based sources as a significant contributor to MPs in aquatic ecosystems. It includes garbage dumping, littering, laundry, tourism, plastic manufacturing and recycling industries, sewage treatment plants, agricultural discharge, recreational activities, fishing, road and tire wear particles, atmospheric deposition, etc. (Baensch-Baltruschat et al., 2020; Vanapalli et al., 2021). The flow velocity, residence time of water, population density, and waste management practices in the area are the crucial factors that influence the abundance and distribution of MPs in the sediments of water bodies. It was observed that there is no unified standard protocol for collecting, processing, and extracting MPs from water and sediment samples. Moreover, the size range of MPs reported often varied significantly across studies. Therefore, comparing the MPs across different studies is difficult. The abundance of MPs reported in rivers and lakes might be an underestimation as the methodologies adopted by most of the authors did not account for smaller MPs.

India currently has 75 Ramsar sites (wetlands of international importance) with a surface area of 1,326,677 hectares (PIB, 2022). However, the MP contamination

Table 1. MPs in Indian Lakes and Rivers.

Sl No.	Location	Sample Type	Abundance of MPs	Size range of MPs	Shape of MPs	Identified Polymers*	References
1	Lake Vembanad, Kerala	Sediment	252.80 ± 25.76 particles per m^2	< 5 mm	Film, foam, fiber, pellet, andfragment	LDPE, PS, PP, and HDPE	(Sruthy and Ramasamy, 2017)
2	LakeRed Hills, Chennai, Tamil Nadu	Sediment	27 particles per kg	0.3 to 5 mm	Fiber, fragment, film, and pellet	HDPE, LDPE, PP, and PS	Gopinath et al., 2020
3	Lake Veeranam, Tamil Nadu	Sediment	309 particles per kg	0.3 to 5 mm	Fiber, film, pellet,and fragment	Nylon, PE, PS, PP, and PVC	(Bharath et al., 2021)
4	Lake Renuka, Himachal Pradesh	Sediment	15–632 particles per kg	< 4.75 mm	Fiber, fragment, foam, and film	PE, PP, and PS	(Ajay et al., 2021)
5	LakeRed Hills, Chennai, Tamil Nadu	Water	5.9 particles per liter	0.3 to 5 mm	Fiber, fragment, film,and pellet	HDPE, LDPE, PP, and PS	Gopinath et al., 2020
6	Lake Veeranam, Tamil Nadu	Water	28 particles per km^2	0.3 to 5 mm	Fiber, film, pellet, and fragment	Nylon, PE, PS, PP, andPVC	(Bharath et al., 2021)
7	Lake Renuka, Himachal Pradesh	Water	02–64 particles per liter	> 0.2 µm	Fiber, fragment, foam, film, and bead/pellet	PE and PP	(Ajay et al., 2021)
8	Ganga River	Sediment	107.57–409.86 particles per kg	63 µmto 10 mm	Fiber, filament, film, foam/ bead,and fragment	PET, PE, PP, and PS	(Sarkar et al., 2019)
9	Netravathi River	Sediment	9.44–253.27 pieces per kg	0.3 to 5 mm	Fragment, fiber, film, foam, and pellet	PE, PET, PP, and PVC	(Amrutha and Warrier, 2020)
10	Brahmaputra River	Sediment	20 to 3,485 particles per kg	20 µm to 5 mm	Fragment, fiber, and bead	PP, PE, PA, PET, PVC, and PTFE	(Tsering et al., 2021)
11	Indus River	Sediment	60 to 1,752 particles per kg	20 µmto 5 mm	Fragment andfiber	PP, PE, PET, PA, PTFE, and PS	(Tsering et al., 2021)
12	Ganga River	Sediment	17–36 particles per kg	0.7 to 7.5 mm	Fragment, foam, film, andfilament	PE, PS, PP, PVC, PE-P, Cellophane, and Polyester	(Singh et al., 2021)
13	Netravathi River	Water	56–2328 particles per m^3	0.3 to 5 mm	Fiber, film, fragment, and foam, pellet	PE, PET, PP, and PVC	(Amrutha and Warrier, 2020)
14	Ganga River	Water	0.038 particles per liter	< 300 µm	Fiber and fragment	Rayon, acrylic, PET, PVC, polyester, and nylon	(Napper et al., 2021)
15	Ganga River	Water	380–684 particles per 1000 m^3	0.7 to 7.5 mm	Fragment, foam, film, and filament	PE, PS, PVC, PE-P, Polyester, and Cellophane	(Singh et al., 2021)

*PE – Polyethylene; PS – Polystyrene; PP – Polypropylene; PVC – Poly Vinyl Chloride; PA – Polyamide; LDPE – Low Density Poly Ethylene; HDPE – High Density Poly Ethylene; PTFE – Polytetrafluoroethylene; PET: Polyethylene terephthalate; PE – P: Polyethylene Propylene in a few Ramsar sites, such as Vembanad and Renuka lakes, has been studied so far. Similarly, very few rivers are studied for their MPs' contamination. Most of the studies are based on one-time sampling, which may not be adequate to understand the temporal variations, sources, and dynamics of MPs in the environment. Moreover, there is a need for unified sampling, extraction, and analysis protocols for MP research in India. The size of the MPs monitored should be unified and smaller particles need to be accounted for considering their greater impact on human health as compared to larger particles. Therefore, nationwide comprehensive monitoring programs using standard protocols are required in the future to understand the spatial and temporal variations in MPs.

MPs in the Marine Environment and Beaches

The marine environment is considered a major sink of MPs. India has a long coastline of 7,500 km, which makes the country one of the world's major fishing and shipping hubs. The coastal regions are densely populated, and several major cities are located on the coast. The urban runoff, sewage, industrial effluents, and discharge from numerous rivers into the coastal waters make it a hot spot for MP pollution.

The beach sediments, beach sand, and surface water have been examined for MP contamination from various coastal regions in India. The MPs reported from the marine environment are diverse in their morphology and polymer composition. MPs are commonly classified as fibers, films, fragments, and pellets. For identifying the polymer content of MPs, most of the studies used FTIR. However, few studies adopted Raman spectroscopy and SEM EDX methods. The polymers, like PE and PP, are the most abundant polymers reported from the marine environment. However, other polymers, such as PS, PET, PVC, PA, PMMA, etc., were also reported in less abundance (Table 2). The time of sample collection, method of sample collection and extraction of MPs, reagents used for density separation, and techniques used for polymer identification influenced the abundance of MPs. Due to the variations in these parameters among the studies reported, a comparison of spatial and temporal variations in MPs is not feasible (Veerasingam et al., 2020).

A threefold increase in MPs in the coastal environment was observed after a flood event in Chennai, Tamil Nadu (Veerasingam et al., 2016). They also observed that the MP abundance is higher near the river discharge points. A similar observation was also reported in Kerala (Kumar and Varghese, 2021a). Therefore, it can be inferred that floods can transport large quantities of plastic debris, including MPs from land-based sources, into the oceans. The abundance of MPs in the coastal environment also may be influenced by the direction of winds and currents (Veerasingam et al., 2016).

Identifying the sources and routes is a critical primary step in the regulation of MP pollution (Vanapalli et al., 2021). The reported studies trace the source of the MPs in the marine environment to land-based sources. The major land-based sources include improper solid waste management, discharge of treated/untreated sewage, inadequate wastewater treatment, industrial effluents, etc. From these sources, the MPs enter the marine environment through stormwater runoff, rivers, and streams. Shipping, fishing, tourism, and pilgrimage activities also lead to the discharge of MPs into the coastal environments (Kumar and Varghese, 2021b, 2021a; Patchaiyappan et al., 2020; Vanapalli et al., 2021).

Table 2. MPs in the Marine Environment and Beaches in India.

Sl No.	Location	Sample Type	Abundance of MPs	Size Range of MPs	Shape of MPs	Commonly Identified Polymers*	References
1	Chennai, Tamil Nadu	Beach sediment	304 (before the flood) to 896 (after the flood)	2 to 5 mm	Pellets	PE and PP	(Veerasingam et al., 2016)
2	Tamil Nadu	Beach sediment	2 to 178 particles per m^2	0.3 to 4.75 mm	Fragment, fiber, and foam	PE, PP, and PS	(Karthik et al., 2018)
3	Tuticorin, Tamil Nadu	Sediment	8.22 ± 0.92 to 17.28 ± 2.53 particles per kg	0.005 mm to 5 mm	Fiber, fragment, and film	PE, PP, PA, polyester,and paint	(Patterson et al., 2019)
4	Port Blair Bay, Andaman Islands	Sediment	45.17 ± 25.23 particles per kg	46.72 to 5024 µm	Fiber, fragment, and pellet	Nylon, PU, PVC, and Acrylic	(Goswami et al., 2020)
5	South Andaman	Beach sediment	414.35 ± 87.4 particles per kg	< 5 mm	Fragment, fiber, and spherule	PP, PVC, nylon, polyvinyl formal, and polybutadiene	(Patchaiyappan et al., 2020)
6	Maharashtra, Karnataka, Goa	Beach sand	162 to 820 particles per m^2	1–5 mm	Fragment, fiber, film, and pellet	PE and PP	(Maharana et al., 2020)
7	Mandovi-Zuari Estuary, Goa	Sediment	800 to 17,300 particles per kg.(Average 4873 to 7,314 particles per kg)	20 µm to 5 mm	Fragment, film, fiber, and bead	PA, PVP, PVC, PAM, polyacetylene, polyimide (PI)	(Gupta et al., 2021)
8	Calicut Beach, Kerala	Beach sediment	80.56 to 467.13 particles per kg	1–5 mm	Film, fiber, and lump,	PE, PP, PET, PE + PP, PS, PVC, and PCU	(Kumar and Varghese, 2021a)
9	Tuticorin coast, Gulf of Mannar	Water	12.14 ± 3.11 to 31.05 ± 2.12 particlesper liter	0.005 to 5 mm	Fiber, fragment, and film	PE and PP	(Patterson et al., 2019)
10	Port Blair Bay, Andaman Islands	Surface seawater	0.93 ± 0.59 particles per m^3	35.29 to 5,010 µm	Fiber, fragment, and pellet	Nylon, PU, PVC, and acrylic	(Goswami et al., 2020)
11	Mandovi-Zuari Estuary, Goa	Water	0.057 to 0.141 particles per m^3	20 µm to 5 mm	Fragment, film, fiber, andbeads	PA, PVP, PVC, PAM, polyacetylene, and polyimide (PI)	(Gupta et al., 2021)
12	Northern coast of Kerala	Surface seawater	0.96–7.12 particles per m^3	0.08 to 2 mm	Fiber, fragment, and flake	PP, PEVA, PA, PE, and polybutadiene	(Pavithran, 2021)

*PE – Polyethylene; PP – Polypropylene; PS – Polystyrene; PA – Polyamide; PVC – Polyvinyl chloride; PAM – Polyacrylamide; PET – Polyethylene terephthalate; PAN – Polyacrylonitrile; PMMA – Polymethyl methacrylate; PEVA – Polyvinyl acetate ethylene; PVAL – Polyvinyl alcohol; PVP – Polyvinyl pyrrolidone; PCU –Polycarbonate urethane.

MPs in Salt

The widespread occurrence of MPs in the marine ecosystem, including water, sediment, and biota, is a major concern for human health. The high abundance of MPs reported in coastal water indicates the possible transfer of these particles to the human food chain via salts. India has a long coastal area and is one of the major salt-producing countries in the world. The leading salt-producing states in India are Gujarat and Tamil Nadu (Vidyasakar et al., 2021). Evaporation of coastal water in salt pans is the primary method of salt production in India. Hence, contaminated coastal water may be a significant source of MPs in salts. The review of reported literature on MPs in salts from India revealed that all the samples irrespective of their origin are contaminated with MPs (Table 3). The abundance of MPs in salts ranges from 2 particles per kg (Sathish et al., 2020a) to 575 particles per kg (Vidyasakar et al., 2021). The size range selected for studying MPs varies considerably from 3.8 μm to 5.2 mm between studies and might influence the abundance of MPs reported. The common shapes of MPs found in salts include fibers and fragments; few studies also reported the presence of films and pellets. FTIR was the most commonly used technique for identifying polymers from salt samples. The different polymers reported are PE, PS, PA, PET, PVC, Nylon, Polyester, etc. (Table 3). The studies suggest that contaminated seawater is the primary source of MPs in salts. However, the possibility of contamination during the processing and packaging of salts cannot be ruled out.

A comparative study of salts produced from Gujarat and Tamil Nadu suggests that MP abundance is high in salts produced from Gujarat than from Tamil Nadu (Vidyasakar et al., 2021). The average concentration of MPs (34 particles per kg salt) from commercially available six brands of sea salts in the state of Kerala has been reported by Ramasamy et al. (2019). By comparing salts produced from different sources, sea salts are more contaminated with MPs than bore-well salts (Sathish et al., 2020a). They estimated that people using sea salts consume 216 particles of MPs per year, while people consuming bore well salts consume 48 particles per year. It was also suggested that a simple sand filtration of seawater before evaporation could significantly reduce the MP content in salt (Seth and Shriwastav, 2018).

MPs in Dust

The MP contamination in the terrestrial environments in India has received little attention compared to the aquatic ecosystems. The MPs in street dust from Chennai, Tamil Nadu, have been reported to contain 227.94 ± 91.37 MPs per 100 g, and the particles are mostly fragments and fibers (Patchaiyappan et al., 2021). The nine polymers identified in the street dust are PVC, HDPE, poly(ethylene-co-vinyl-acetate), poly (tetrafluoroethylene), cellulose microcrystalline, lyocell, superflex-200, wax-1032, and AC-395 (Patchaiyappan et al., 2021). On the other hand, indoor dust from Patna, India, has been reported to contain MPs, such as PET (55–6,800 μg/g) and PC (0.11–530 μg/g) (Zhang et al., 2020). The above mentioned studies signify the importance of dust as a route of exposure to MPs through inhalation. The sources of MPs in the indoor dust may be attributed to textile fibers, furniture, packaging items, toys, carpets, paints, etc. The MPs in street dust may be attributed to improper waste management, littering, tire and road wear particles, road paints, building materials, and atmospheric transport.

Table 3. MPs in Salt.

Sl No.	Location	Sample Type	Abundance of MPs	Size range of MPs	Shape of MPs	Identified Polymers*	References
1	Kerala, Maharashtra, Gujarat	Commercial salt	56 ± 49 to 103 ± 39 particles per kg	< 5 mm	Fiber and fragment	PE, PS, PET, PA, and polyesters	(Seth and Shriwastav, 2018)
2.	Kerala	Commercial salt	34 particles per kg	< 5 mm	Fiber, fragment, bead, and sheet	PS, PA, PET, PP, PE, and PVC	Ramasamy et al. (2019)
3	Tamil Nadu	Salt pans	Na	na	Fragment, fiber, and film	PP, PE, nylon, and cellulose	(Selvam et al., 2020)
4	Tamil Nadu	Sea salt	35 ± 15 to 72 ± 40 particles per kg	55 μm to 2 mm	Fiber and fragment	PE, PP, Polyester, and PA	(Sathish et al., 2020a)
5	Tamil Nadu	Tube welll salt	2 ± 1 to 29 ± 11 particles per kg	55 μm to 2 mm	Fiber and fragment	PE, PP, Polyester, and PA	(Sathish et al., 2020a)
6	Tamil Nadu	Salt	23 to 101 particles per 200 g	100 μm to 5mm	Fiber, film, pellet, and irregular	PE, Polyester, and PVC	(Vidyasakar et al., 2021)
7	Gujarat	Salt	46 to 115 particles per 200 g	100 μm to 5mm	Fiber, film, pellet, and irregular	PE, Polyester, and PVC	(Vidyasakar et al., 2021)
8	Tamil Nadu	Salt from salt pans	3.67 ± 1.54 to 21.33 ± 1.53 particles per 10g	na	Fiber	Nylon, PP, LDPE, and PET	(Nithin et al., 2021)
9	Tamil Nadu	Commercial salt	4.67 ± 1.15 to 16.33 ± 1.53 particles per 10g	na	Fiber	Nylon, PP, LDPE, and PET	(Nithin et al., 2021)
10	India	Commercial salt	700 particles per kg	3.8 μm to 5.2 mm	Fragment, fiber, and pellet	Cellophane, PS, PA, and polyarylether (PAR)	(Sivagami et al. 2021)

*PE – Polyethylene; PS – Polystyrene; PP – Polypropylene; PVC – Poly Vinyl Chloride; LDPE – Low Density Poly Ethylene; PA – polyamide; PVC – Polyvinyl chloride.

MPs in Biota

The infiltration of MPs in the human food chain is drawing more attention globally. In this context, the presence of MPs in different aquatic organisms, such as zooplanktons, shrimps, mussels, oysters, and fishes, has been reported from various parts of India (Table 4). Most of the studies came from south Indian states, like Kerala and Tamil Nadu. In the extraction procedure of MPs, the majority of the studies used 10% KOH for the digestion of organic matter. However, very few studies used H_2O_2 for the removal of organic matter. Fibers, fragments, films, and pellets are commonly reported in biota. Few studies also reported beads and foam. FTIR was most commonly used to identify the polymers. Very few studies used Raman spectroscopy for polymer identification. A number of investigators used SEM EDAX to evaluate the surface morphology and estimate the adsorbed chemicals, including heavy metals, on the surface of MPs. Polymers, such as PE, PP, PS, PVC, nylon, and acrylic, are most commonly reported to be ingested by aquatic species. The abundance of MPs varies considerably depending on the species, feeding behavior, size, and age of the organism investigated. MPs were observed to be more in pelagic fishes than demersal fishes (James et al., 2021). We observed that most of the studies reported a high abundance of low-density polymers, such as PE and PP in the surface waters. Hence, it could be inferred that the pelagic fishes are exposed to higher levels of low-density MPs that float in the seawater. However, more studies are required to reach a consensus on this matter.

Few studies investigated the toxic effects of MPs on biota. It is reported that MPs cause abrasion in the ciliated structure and induced toxic physiological and structural alterations in the mussel *Pernaviridis* (Vasanthi et al., 2021). They also reported significantly high levels of oxidative stress markers in response to MP exposure in *P. viridis*. An increase in lipid peroxidation and enzymatic antioxidants in the shrimp *Litopenaeus vannamei* due to short-term exposure to PE microbeads was also reported (Maharana et al., 2020). These studies highlight the need to investigate the ecological implications of MP pollution.

Most of the studies investigated MPs in the digestive tracts of fishes. However, except for small fishes, the digestive tract is normally removed before cooking. However, dietary exposure to MPs via fish could be possible if these particles were able to translocate across the gastrointestinal tract (GIT) or gills via paracellular diffusion or transcellular uptake (Wright and Kelly, 2017). On the other hand, plastics contain several additives, such as phthalates, bisphenols, colorants, etc., which are known as endocrine disruptors. Moreover, owing to their large surface-to-volume ratio, MPs can accumulate significantly high amounts of pollutants from the surrounding environment and can act as vectors of these pollutants to the organisms. The release of such additives and adsorbed pollutants in the digestive system of fishes may adversely affect the organism concerned. Additionally, the accumulation of these pollutants in the tissues of such organisms assumes a greater risk to human health and therefore needs more attention in the future.

MPs' sources, routes of exposure, and human health impacts

MPs are ubiquitous pollutants in the environment and have been reported from a number of ecosystems and food items destined for human consumption. MPs are also reported to be ingested by fishes and other aquatic organisms. Therefore, human exposure to MPs is unavoidable. The sources, routes of exposure, and possible health

Table 4. MPs in the Aquatic Biota.

Sl. No.	Type of Biota	Location	Abundanceof MPs	Size Range of MPs	Shape of MPs	Polymer Type*	References
1	Zooplankton	Port Blair Bay, Andaman Islands	0–0.41 particles per individual	21.57 to 2,225 μm	Fiber, fragment, pellet	Nylon, PU, PVC, and Acrylic,	(Goswami et al., 2020)
2	Shrimps	Kochi, Kerala	0.39 ± 0.6 particles per shrimp	157 to 2,785 μm	Fiber, fragment, sheet	PE, PP, PA, and Polyester	(Daniel et al., 2020b)
3	Mussels	Kasimedu, Chennai	-	0.8 mm to 4.7 mm	Fiber, sphere, flake, sheet, fragment	PP, PE, polyester, cellophane, rayon	(Vasanthi et al., 2021)
4	Oyster	Tuticorin coast, Gulf of Mannar	6.9 ± 3.84 particles per individual	0.005–5 mm	Fiber, film, fragment	PE, and PP	(Patterson et al., 2019)
5	Fishes and shellfishes	Port Blair Bay, Andaman Islands	10.65 ±7.83 particles per specimen	111.58 to 5094 μm	Fiber, fragment, and pellet	Nylon, PU, PVC, and acrylic	(Goswami et al., 2020)
6	Fishes	Chennai and Nagapattinam of Tamil Nadu	20 particles in 17 fishes	< 5 mm	Fiber, film, pellet	PE, PA, and polyester	(Karuppasamy et al., 2020)
7	Fishes	Vembanad Lake, Kerala	15 ± 13 particles per fish	0.04–4.73 mm	Fiber, fragment, film, microbead	PE, PP, PS, and PVC	(Nikki et al., 2021)
8	Fishes	Tuticorin, Tamil Nadu	0.11 ± 0.06 to 3.64 ± 1.7 particles per individual	85 μm to 5 mm	Fiber, fragment, film, foam	PE, PA, and polyester	(Sathish et al., 2020b)
9	Fishes	Kerala	22 particles from 15 fishes	< 5 mm	Fiber, fragment, foam	PE, cellulose, rayon, polyester, and PP	(Robin et al., 2020)
10	Fishes	Kochi, Kerala	30 particles from 653 samples	0.27 mm to 3.2 mm	Fragment, filament, pellet	PE and PP	(James et al., 2020)

*PE – Polyethylene; PS – Polystyrene; PP – Polypropylene; PVC – Poly Vinyl Chloride; PA – Polyamide; PU – Polyurethane.

impacts of MPs on humans are summarized in Fig. 1. Cox et al. (2019) estimated that annual MPs consumption by humans through food and beverages ranged from 39,000 to 52,000 particles, depending on age and sex. These numbers may increase significantly when exposure through drinking water and inhalation is considered. MPs less than 130 μm in diameter have the potential to translocate into human tissues and initiate a localized immune response (Wright and Kelly, 2017). Inside the tissues, they may also release constituent monomers, additives, and adsorbed contaminants, such as persistent organic pollutants and heavy metals (Cox et al., 2019; Wright and Kelly, 2017). The recent findings of MPs in the feces of humans (Schwabl et al., 2019; Zhang et al., 2021) and the human placenta (Ragusa et al., 2021) highlighted the consumption of MPs by humans and its possible health consequences. However, further studies are required to shed more light on the human health impacts of MPs.

Policies in India

In order to provide a regulatory framework for the systematic management of plastic waste in the country, the government of India has issued several rules and amendments from time to time. These include the Plastic Waste (Management and Handling) Rules (2011), Plastic Waste Management Rules (2016), Plastic Waste Management (Amendment) Rules (2018), and more recently the Plastic Waste Management (Amendment) Rules (2021). India has also made significant interventions in the international forums for mitigating plastic pollution. In 2019, the United Nations Environment Assembly adopted a resolution piloted by India to address pollution from single-use plastic (SUP) products.

These above mentioned rules mandated the generators of plastic waste take necessary measures that minimize the quantity and littering of plastic waste and store segregated plastic waste at the source. These rules stipulated the responsibilities of waste generators, local bodies, gram panchayats, street vendors, and retailers to effectively manage plastic waste. The rules mandated Extended Producer Responsibility (EPR) by the producers, importers, and brand owners and worked out modalities for collecting plastic waste. State-level monitoring committees have been constituted under these rules to effectively monitor and implement the provisions of these rules. Through the Plastic Waste Management (Amendment) Rules, 2021, the government of India aims to prohibit the production and use of identified single-use plastic items having low utility and high littering potential by 2022. Accordingly, the minimum thickness of plastic carry bags was mandated to increase from 50 microns to 75 microns on September 30, 2021. It has been further mandated to increase to 120 microns by December 31, 2022. Moreover, the government prohibited the manufacturing, importing, stocking, distributing, selling,and using of single-use plastic products, such as plastic plates, cups, glasses, cutleries, wrapping or packing films around sweet boxes, invitation cards, cigarette packets, plastic or PVC banners less than 100 microns in thickness,plastics sticks for earbuds, candy, and ice cream, balloons, and plastic flags with effect from July 1, 2022. The waste management rules, along with the budget allocation of Rs. 1,41,678 crore over 5 years from 2021–2026 under the Urban Swachh Bharat Mission 2.0, is expected to improve the environment and health through complete fecal sludge management and wastewater treatment, source segregation of garbage, reduction in single-use plastic, reduction in air pollution by effectively managing waste from construction-and-demolition activities, and bio-remediation of all legacy dump sites.

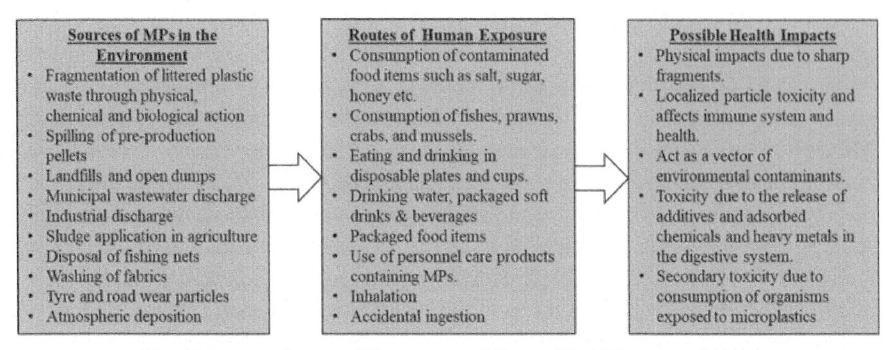

Fig. 1. Sources, Routes of Exposure, and Human Health Impacts of MPs.

The above mentioned rules and policies are certainly an impetus for reducing plastic waste; however, considering the gravity of the problem, more stringent measures are required. We believe that public action is the deciding factor for reducing plastic and MP pollution. Therefore, along with the stringent implementation of the rules, we propose the following measures for better reduction of plastic waste and MP pollution.

- Systematic and progressive ban on single-use plastic products and promote environment-friendly alternate materials.
- Ban the use of MPs in paints, personal care products, cleaning products, etc.
- Improve infrastructure for plastic waste recycling.
- Promote circular economy and ensure strict compliance to extended producer responsibility.
- Sensitize the issues of plastics and MPs through regular public awareness campaigns for civilians through mass media and students'curriculum.
- Create awareness among fishermen and give incentives for measures taken to reduce plastic pollution.
- Regular cleanup drives with public participation to remove littered plastics from various environments, especially beaches.
- Impose taxes on plastics and provide fiscal incentives to reduce plastic.
- Popularize the concept of plastic-free markets, campuses, offices, etc.
- Recognize and adopt best practices for plastic waste management.
- Improve waste management and ensure zero dumping of waste in the open.
- Improve wastewater and sludge management.

Future Research Perspectives

Since most of the studies on MPs are conducted in South India, there is a paucity of data on the occurrence and abundance of MPs from other regions of the country. There is also a need to standardize sample collection and analysis protocol, which will help to compare results and evaluate the spatial and temporal variations in MP pollution. Since MPs are present in large numbers in sediments, the chances of further fragmentation of these particles cannot be ruled out. The long residence time of MPs in sediments may lead to their vertical migration in the sediment column. This has further implications for the contamination of groundwater aquifers. Therefore, we suggest that future research should also focus on the vertical distribution of MPs in sediments through core samples. Most MPs reported in the studies belong to secondary MPs. The sources of them are the fragmentation of larger plastics in the environment. Therefore, future studies need to be focused on tracing the sources of MPs for better control of MP pollution.

MPs in the terrestrial ecosystems, atmosphere, and indoor environments received little attention in India. Therefore, more research needs to be focused on these environments and the exposure of terrestrial organisms and humans to MPs. Research should also focus on developing and improving the waste and wastewater treatment processes to reduce MPs. Apart from this, the impact of policies on the abundance and distribution of MPs needs to be monitored systematically.

Acknowledgments

The authors are grateful to the Ministry of Earth Sciences (MoES), Govt. of India, for funding the project "Microplastics Contamination in Sediment and Selected Benthic Organisms of Arabian Sea Coast and a Few Estuaries of Kerala" through National Centre for Coastal Research (NCCR), Chennai. The financial support from the Kerala State Council for Science Technology and Environment (KSCSTE), Govt. of Kerala, for funding the project "Monitoring and Assessment of Microplastic Pollution in Water, Sediment and Selected Biota of Vembanad Lake, Kerala" is gratefully acknowledged as well. Lastly, the authors acknowledge DST-SAIF and DST-PURSE (Phase II) programs of Mahatma Gandhi University for providing the infrastructure and analytical facilities for this research.

References

Ajay, K., Behera, D., Bhattacharya, S., Mishra, P.K., Ankit, Y. and Anoop, A. 2021. Distribution and characteristics of microplastics and phthalate esters from a freshwater lake system in Lesser Himalayas. Chemosphere 283: 131132.

Amrutha, K. and Warrier, A.K. 2020. The first report on the source-to-sink characterization of microplastic pollution from a riverine environment in tropical India. Sci. Total Environ. 739: 140377.

Andrady, A.L. 2011. Microplastics in the marine environment. Mar. Pollut. Bull. 62: 1596–1605.

Baensch-Baltruschat, B., Kocher, B., Stock, F. and Reifferscheid, G. 2020. Tyre and road wear particles (TRWP) - A review of generation, properties, emissions, human health risk, ecotoxicity, and fate in the environment. Sci. Total Environ. 733: 137823.

Beriot, N., Peek, J., Zornoza, R., Geissen, V. and Huerta Lwanga, E. 2021. Low density-microplastics detected in sheep faeces and soil: A case study from the intensive vegetable farming in Southeast Spain. Sci. Total Environ. 755: 142653.

Bharath K.M., Srinivasalu, S., Natesan, U., Ayyamperumal, R., Kalam S, N., Anbalagan, S., Sujatha, K. and Alagarasan, C. 2021. Microplastics as an emerging threat to the freshwater ecosystems of Veeranam lake in south India: A multidimensional approach. Chemosphere 264: 128502.

Chen, G., Feng, Q. and Wang, J. 2020. Mini-review of microplastics in the atmosphere and their risks to humans. Sci. Total Environ. 703: 135504.

Cox, K.D., Covernton, G.A., Davies, H.L., Dower, J.F., Juanes, F. and Dudas, S.E. 2019. Human Consumption of Microplastics. Environ. Sci. Technol. 53: 7068–7074.

CPCB, 2019. Annual report on implementation of plastic waste management rules, 2018 for the year 2018-2019. Central Pollution Control Board, Ministry of Environment Forest and Climate Change, Government of India, October 2019.

Daniel, D.B., Ashraf, P.M. and Thomas, S.N. 2020a. Microplastics in the edible and inedible tissues of pelagic fishes sold for human consumption in Kerala, India. Environ. Pollut. 266: 115365.

Daniel, D.B., Ashraf, P.M. and Thomas, S.N. 2020b. Abundance, characteristics and seasonal variation of microplastics in Indian white shrimps (Fenneropenaeus indicus) from coastal waters off Cochin, Kerala, India. Sci. Total Environ. 737: 139839.

de Souza Machado, A.A., Kloas, W., Zarfl, C., Hempel, S. and Rillig, M.C. 2018. Microplastics as an emerging threat to terrestrial ecosystems. Glob. Chang. Biol. 24: 1405–1416.

Dong, M., Luo, Z., Jiang, Q., Xing, X., Zhang, Q. and Sun, Y. 2020. The rapid increases in microplastics in urban lake sediments. Sci. Rep. 10: 1–10.

Geyer, R., Jambeck, J.R. and Law, K.L. 2017. Production, use, and fate of all plastics ever made. Sci. Adv. 3: 25–29.

González-Pleiter, M., Velázquez, D., Edo, C., Carretero, O., Gago, J., Barón-Sola, Á., Hernández, L.E., Yousef, I., Quesada, A., Leganés, F., Rosal, R. and Fernández-Piñas, F. 2020. Fibers spreading worldwide: Microplastics and other anthropogenic litter in an Arctic freshwater lake. Sci. Total Environ. 722: 137904.

Gopinath, K., Seshachalam, S., Neelavannan, K., Anburaj, V., Rachel, M., Ravi, S., Bharath, M. and Achyuthan, H. 2020. Quantification of microplastic in Red Hills Lake of Chennai city, Tamil Nadu, India. Environ. Sci. Pollut. Res. 27: 33297–33306.

Goswami, P., Vinithkumar, N.V. and Dharani, G. 2020. First evidence of microplastics bioaccumulation by marine organisms in the Port Blair Bay, Andaman Islands. Mar. Pollut. Bull. 155: 111163.

Gupta, P., Saha, M., Rathore, C., Suneel, V., Ray, D., Naik, A., Unnikrishnan, K., Divya M. and Daga, K. 2021. Spatial and seasonal variation of microplastics and possible sources in the estuarine system from central west coast of India. Environ. Pollut. 288: 117665.

Jambeck, J.R., Geyer, R., Wilcox, C., Siegler, T.R., Perryman, M., Andrady, A., Narayan, R. and Law, K.L. 2015. Plastic waste inputs from land into the ocean. Science. 347: 768 –771.

James, K., Vasant, K., Padua, S., Gopinath, V., Abilash, K.S., Jeyabaskaran, R., Babu, A. and John, S. 2020. An assessment of microplastics in the ecosystem and selected commercially important fishes off Kochi, south eastern Arabian Sea, India. Mar. Pollut. Bull. 154: 111027.

James, K., Vasant, K., Sikkander Batcha, S.M., Padua, S., Jeyabaskaran, R., Thirumalaiselvan, S., Vineetha, G. and Benjamin, L.V. 2021. Seasonal variability in the distribution of microplastics in the coastal ecosystems and in some commercially important fishes of the Gulf of Mannar and Palk Bay, Southeast coast of India. Reg. Stud. Mar. Sci. 41: 101558.

Karthik, R., Robin, R.S., Purvaja, R., Ganguly, D., Anandavelu, I., Raghuraman, R., Hariharan, G., Ramakrishna, A. and Ramesh, R. 2018. Microplastics along the beaches of southeast coast of India. Sci. Total Environ. 645: 1388–1399.

Karuppasamy, P.K., Ravi, A., Vasudevan, L., Elangovan, M.P., Dyana Mary, P., Vincent, S.G.T. and Palanisami, T. 2020. Baseline survey of micro and mesoplastics in the gastro-intestinal tract of commercial fish from Southeast coast of the Bay of Bengal. Mar. Pollut. Bull. 153: 110974.

Kumar, A.S. and Varghese, G.K. 2021a. Microplastic pollution of Calicut beach - Contributing factors and possible impacts. Mar. Pollut. Bull. 169: 112492.

Kumar, A.S. and Varghese, G.K. 2021b. Source Apportionment of Marine Microplastics: First Step Towards Managing Microplastic Pollution. Chem. Eng. Technol. 44: 906–912.

Kutralam-Muniasamy, G., Pérez-Guevara, F., Elizalde-Martínez, I. and Shruti, V.C. 2020. Branded milks – Are they immune from microplastics contamination? Sci. Total Environ. 714: 136823.

Lebreton, L.C.M., Van Der Zwet, J., Damsteeg, J.W., Slat, B., Andrady, A. and Reisser, J. 2017. River plastic emissions to the world's oceans. Nat. Commun. 8: 1–10.

Maharana, D., Saha, M., Dar, J.Y., Rathore, C., Sreepada, R.A., Xu, X.R., Koongolla, J.B. and Li, H.X. 2020. Assessment of micro and macroplastics along the west coast of India: Abundance, distribution, polymer type and toxicity. Chemosphere 246: 125708.

Mani, T., Hauk, A., Walter, U. and Burkhardt-Holm, P. 2015. Microplastics profile along the Rhine River. Sci. Rep. 5: 17988.

Napper, I.E., Baroth, A., Barrett, A.C., Bhola, S., Chowdhury, G.W., Davies, B.F.R., Duncan, E.M., Kumar, S., Nelms, S.E., Hasan Niloy, M.N., Nishat, B., Maddalene, T., Thompson, R.C. and Koldewey, H. 2021. The abundance and characteristics of microplastics in surface water in the transboundary Ganges River. Environ. Pollut. 274: 116348.

Nikki, R., Abdul Jaleel, K.U., Abdul Ragesh, S., Shini, S., Saha, M. and Dinesh Kumar, P.K. 2021. Abundance and characteristics of microplastics in commercially important bottom dwelling finfishes and shellfish of the Vembanad Lake, India. Mar. Pollut. Bull. 172: 112803.

Nithin, A., Sundaramanickam, A., Surya, P., Sathish, M., Soundharapandiyan, B. and Balachandar, K. 2021. Microplastic contamination in salt pans and commercial salts—A baseline study on the salt pans of Marakkanam and Parangipettai, Tamil Nadu, India. Mar. Pollut. Bull. 165: 112101.

Patchaiyappan, A., Ahmed, S.Z., Dowarah, K., Jayakumar, S. and Devipriya, S.P. 2020. Occurrence, distribution and composition of microplastics in the sediments of South Andaman beaches. Mar. Pollut. Bull. 156: 111227.

Patchaiyappan, A., Dowarah, K., Zaki Ahmed, S., Prabakaran, M., Jayakumar, S., Thirunavukkarasu, C. and Devipriya, S.P. 2021. Prevalence and characteristics of microplastics present in the street dust collected from Chennai metropolitan city, India. Chemosphere 269: 128757.

Patterson, J., Jeyasanta, K.I., Sathish, N., Booth, A.M. and Edward, J.K.P. 2019. Profiling microplastics in the Indian edible oyster, Magallana bilineata collected from the Tuticorin coast, Gulf of Mannar, Southeastern India. Sci. Total Environ. 691: 727–735.

Pavithran, V.A. 2021. Study on microplastic pollution in the coastal seawaters of selected regions along the northern coast of Kerala, southwest coast of India. J. Sea Res. 173: 102060.

PIB, 2022. 75 Ramsar sites in 75th year of Independence. Press Information Bureau, Government of India, 13 August 2022.

Ragusa, A., Svelato, A., Santacroce, C., Catalano, P., Notarstefano, V., Carnevali, O., Papa, F., Rongioletti, M.C.A., Baiocco, F., Draghi, S., D'Amore, E., Rinaldo, D., Matta, M. and Giorgini, E. 2021. Plasticenta: First evidence of microplastics in human placenta. Environ. Int. 146: 106274.

Ramasamy, E.V., Sruthy, S., Harit, A.K., Mohan, M. and Binish, M.B. 2021. Microplastic pollution in the surface sediment of Kongsfjorden, Svalbard, Arctic. Mar. Pollut. Bull. 173: 112986.

Ramasamy, E.V., Sruthy, S.N., Harit, A.K., Babu N. Microplastics in human consumption: Table salt contaminated with microplastics. pp. 74-80. In: S. Babel, A Haarstrick, M.S. Babel, A. Sharp [eds.] 2019. Microplastics in the water environment. Cuvillier Verlag, Göttingen.

Robin, R.S., Karthik, R., Purvaja, R., Ganguly, D., Anandavelu, I., Mugilarasan, M. and Ramesh, R. 2020. Holistic assessment of microplastics in various coastal environmental matrices, southwest coast of India. Sci. Total Environ. 703: 134947.

Sarkar, D.J., Das Sarkar, S., Das, B.K., Manna, R.K., Behera, B.K. and Samanta, S. 2019. Spatial distribution of meso and microplastics in the sediments of river Ganga at eastern India. Sci. Total Environ. 694: 133712.

Sathish, M.N., Jeyasanta, I. and Patterson, J. 2020a. Microplastics in Salt of Tuticorin, Southeast Coast of India. Arch. Environ. Contam. Toxicol. 79: 111–121.

Sathish, M.N., Jeyasanta, I. and Patterson, J. 2020b. Occurrence of microplastics in epipelagic and mesopelagic fishes from Tuticorin, Southeast coast of India. Sci. Total Environ. 720: 137614.

Schwabl, P., Koppel, S., Konigshofer, P., Bucsics, T., Trauner, M., Reiberger, T. and Liebmann, B. 2019. Detection of various microplastics in human stool: A prospective case series. Ann. Intern. Med. 171: 453–457.

Selvam, S., Manisha, A., Venkatramanan, S., Chung, S.Y., Paramasivam, C.R. and Singaraja, C. 2020. Microplastic presence in commercial marine sea salts: A baseline study along Tuticorin Coastal salt pan stations, Gulf of Mannar, South India. Mar. Pollut. Bull. 150: 110675.

Seth, C.K. and Shriwastav, A. 2018. Contamination of Indian sea salts with microplastics and a potential prevention strategy. Environ. Sci. Pollut. Res. 25: 30122–30131.

Singh, N., Mondal, A., Bagri, A., Tiwari, E., Khandelwal, N., Monikh, F.A. and Darbha, G.K. 2021. Characteristics and spatial distribution of microplastics in the lower Ganga River water and sediment. Mar. Pollut. Bull. 163: 111960.

Sivagami, M., Selvambigai, M., Devan, U., Velangani, A.A.J., Karmegam, N., Biruntha, M., Arun, A., Kim, W., Govarthanan, M. and Kumar, P. 2021. Extraction of microplastics from commonly used sea salts in India and their toxicological evaluation. Chemosphere 263: 128181.

Sruthy, S. and Ramasamy, E.V. 2017. Microplastic pollution in Vembanad Lake, Kerala, India: The first report of microplastics in lake and estuarine sediments in India. Environ. Pollut. 222: 315–322.

Tsering, T., Sillanpää, Mika, Sillanpää, Markus, Viitala, M. and Reinikainen, S.P. 2021. Microplastics pollution in the Brahmaputra River and the Indus River of the Indian Himalaya. Sci. Total Environ. 789: 147968.

Vanapalli, K.R., Dubey, B.K., Sarmah, A.K. and Bhattacharya, J. 2021. Assessment of microplastic pollution in the aquatic ecosystems – An Indian perspective. Case Stud. Chem. Environ. Eng. 3: 100071.

Vasanthi, R.L., Arulvasu, C., Kumar, P. and Srinivasan, P. 2021. Ingestion of microplastics and its potential for causing structural alterations and oxidative stress in Indian green mussel Perna viridis– A multiple biomarker approach. Chemosphere 283: 130979.

Veerasingam, S., Mugilarasan, M., Venkatachalapathy, R. and Vethamony, P. 2016. Influence of 2015 flood on the distribution and occurrence of microplastic pellets along the Chennai coast, India. Mar. Pollut. Bull. 109: 196–204.

Veerasingam, S., Ranjani, M., Venkatachalapathy, R., Bagaev, A., Mukhanov, V., Litvinyuk, D., Verzhevskaia, L., Guganathan, L. and Vethamony, P. 2020. Microplastics in different environmental compartments in India: Analytical methods, distribution, associated contaminants and research needs. TrAC - Trends Anal. Chem. 133: 116071.

Vidyasakar, A., Krishnakumar, S., Kumar, K.S., Neelavannan, K., Anbalagan, S., Kasilingam, K., Srinivasalu, S., Saravanan, P., Kamaraj, S. and Magesh, N.S. 2021. Microplastic contamination in edible sea salt from the largest salt-producing states of India. Mar. Pollut. Bull. 171: 112728.

Waller, C.L., Griffiths, H.J., Waluda, C.M., Thorpe, S.E., Loaiza, I., Moreno, B., Pacherres, C.O. and Hughes, K.A. 2017. Microplastics in the Antarctic marine system: An emerging area of research. Sci. Total Environ. 598: 220–227.

Wright, S.L. and Kelly, F.J. 2017. Plastic and Human Health: A Micro Issue? Environ. Sci. Technol. 51: 6634–6647.

Zhang, J., Wang, L. and Kannan, K. 2020. Microplastics in house dust from 12 countries and associated human exposure. Environ. Int. 134: 105314.

Zhang, N., Li, Y. Bin, He, H.R., Zhang, J.F. and Ma, G.S. 2021. You are what you eat: Microplastics in the feces of young men living in Beijing. Sci. Total Environ. 767: 144345.

Chapter 12

Forging Plastic Litter Governance

Addressing Acute Plastic Litter Pollution in Korea

Youngjin Choi

INTRODUCTION

It is a fact of life that plastics are an important part of modern life. However, irresponsible disposal of plastic materials is making the planet unsafe for us and our future generations. To present an alternative to plastics, the problem of environmental pollution caused by the reckless use of disposable products on a global level needs to be examined first. Plastic pollution, in particular, provides an acute threat to the ocean, river, and inland lake ecosystems.

In Korea, daily life involves the use of a lot of plastic products, such as straws, cups, and bottles used in drinks, meal boxes, and parcels wrapped in plastic packaging that come through shopping online. Total plastics waste generation in 2019 increased by about 11.5% compared to the previous year, i.e., 446,102 tons/day to 497,238 tons/day. The composition cost for each type of waste in 2019 was 44.5% of construction waste, 40.7% of business discharge facility waste, 11.7% of household waste, and 3.1% of designated waste (Korean Ministry of Environment, 2020).

The trash crisis has been a big concern in Korea for some time now. Korea used to export the waste to China and other countries. However, it started in 2018 when domestic companies refused to collect plastic wastes because China banned the importation of 24 solid waste items due to the high pollution risks they pose to the environment. After this enforcement, environmental activists from the Philippines have likewise expressed opposition to Korea exporting plastic wastes into their country, prompting the Philippine government to return these to Korea twice in the second half of 2018 (KBS News, April 3, 2019).

Department of Sociology and Institute of Global Affairs, Kyung Hee University.

Unit: ton/day

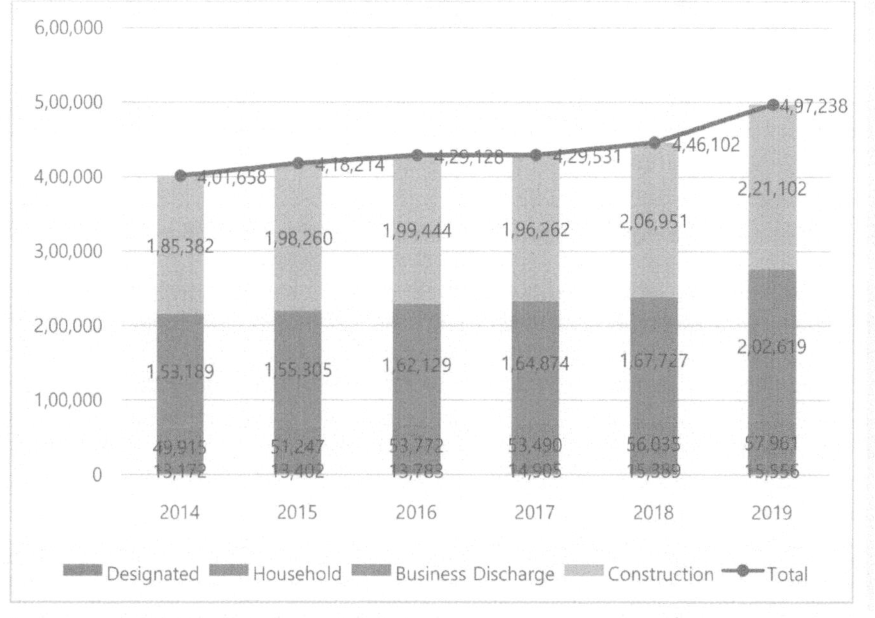

Source: Korean Ministry of Environment (2020)

Fig. 1. The Volume of Plastic Waste Generated by Sort.

Accordingly, the Korean government came up with policies to regulate the use of solid plastics. Unfortunately, the use of disposable plastics is already part of the lives of modern people seeking convenience, and regulating the use of plastics is no longer seen as a fundamental solution (Marine Innovation, 2019). The only option for such is to reuse and recycle them.

This paper discusses the concern about the saturated landfills in the Seoul metropolitan area and the proliferation of waste mountains in its adjoining provinces. The paper also tackles how this is linked to the problems of waste export failure in China and Southeast Asian countries. In addition, the paper examines leaching and cyclical flow of coastal wastes from the inland to the sea of the Korean peninsula. This chapter also attempts to suggest policies to address all the above-mentioned plastic issues such as forging multilayered plastic governance on the use of plastic and recycling in Korea.

Landfill Situation in the Seoul Metropolitan Area

Before 1964, the residential sites, wetlands, etc., of Seoul reeked of bad odor because the city did not have any decent waste disposal plant for solid wastes. Solid wastes were simply dumped and buried anywhere. In 1964, the Seoul municipal government secured landfills on the outskirt of Northern Seoul. To address this situation, the government also converted Gangnam (located on the southern outskirt of the Seoul city center in the north center), which previously used to be rice fields, into an exclusive city landfill. The city used the Gangnam area mainly as a landfill from the mid-1960s to 1977.

During this period, the population of Seoul increased significantly. From 1.56 million in 1955, the city's population increased to 7.82 million in 1978. To address the demands for housing, the landfill sites in the Gangnam area were developed into large-scale apartment complexes, thus putting pressure on the government to consolidate its landfills in one place.

The city of Seoul designated Nanjido, located in the suburban area of Sangam-dong, as a large-scale landfill area. The area was adjudged to be the best place to throw trash far from 'one's neighborhood' and was assigned convenient transportation and levee construction. From March 1978, trash discharged from Seoul and parts of Gyeonggi province were dumped in the Nanjido in Sangam-dong.

Nanjido, near the site of the current World Cup Park, was originally a beautiful island called 'Flower Island'. It was a place where orchids and bushes once grew along flowing clear water. From 1978 to 1993, however, the area was converted into a landfill where all kinds of garbage were dumped and left to accumulate dust and rot, thus emitting a foul odor.

Due to Seoul's rapid expansion and urbanization, garbage trucks lined up to leave all kinds of garbage in Nanjido every day. It was estimated that the average amount of trash generated per person was 1.95 kg in 1985. This increased sharply to 2.17 kg in 1988 and 2.30 kg in 1991. The Seoul Metropolitan Government estimated at that time, an average of 3,000 tons of waste, which was 8-ton trucks a day, were dumped in Nanjido. This continued for 15 years until the Nanjido got filled up in March 1993, and by then it was estimated to contain about 92 million m^3. Household wastes, construction, and industrial wastes were piled randomly in the landfill site, along with two huge piles of garbage that stood at a height of 90 m.

When the Nanjido trash site, which was responsible for holding 78% of Seoul's household wastes, reached its limit in 1993, there was no other place in Seoul that could accommodate the solid wastes of the city anymore. Because Seoul did not

Fig. 2. The Nanjido Landfill in 1985.

want garbage to pile up within the city, the city government then looked for a site outside the city and identified the western reclamation sites of Gimpo-gun, Gyeonggi province, and Oryu-dong, Seo-gu, Incheon as the potential landfill sites. However, the province and the city opposed the idea of Seoul dumping garbage there. To resolve the situation, the National Environment Agency intervened and proposed to create an integrated facility that would treat garbage generated by Seoul, Gyeonggi, and Incheon in one place itself. This facility is now known as the 'Seoul Metropolitan Area Landfill Site'.

The metropolitan landfill site opened on February 10, 1992, just as the Nanjido landfill in Seoul reached its saturation point. The new landfill site was created by landfilling public wastes near Incheon Metropolitan City and Gimpo City in Gyeonggi province. Currently, the said landfill serves as a "wide-area waste treatment facility" that jointly treats wastes generated in Seoul, Gyeonggi-do, and Incheon. This metropolitan landfill site is the world's largest 'garbage bin' with an area of 16.85 m^2 and a reclamation capacity of 230 million tons. The total amount of garbage brought in 2020 was 634,359 tons, while it was 621,953 tons by the end of October, 2020.

Initially, the permit for reclamation in the metropolitan area was the Maginot Line, which was expected to terminate by 2016. However, this has been extended up to August 2025 due to a lack of alternative landfills. As the reclamation period was extended, Seoul, Gyeonggi, and Incheon decided to explore other landfills.

Since the reclaimed land in the metropolitan area has been operating for around 28 years, environmental damage has been accumulating as well, such as increasing odor and leachate. Sawol village, which is about 1 km away from the metropolitan landfill site, has been marked as unsuitable for housing as a consequence of the investigation by the Ministry of Environment. With the construction of Cheongna International City and Geomdan New City, the resistance of residents toward the reclaimed land has also been increasing. This is the reason that around 80% of the total waste carried into the metropolitan landfill is coming from Seoul and Gyeonggi provinces.

Garbage Mountain Problem

In recent years, the problem of domestic waste production has become even more prominent because China, the world's largest importer of plastic waste, declared that it will no longer accept foreign wastes. South Korea attempted to export garbage to Southeast Asian countries, such as the Philippines and Indonesia, as a desperate measure to address this issue. However, discarded plastic wastes that were not collected properly and were not suitable for recycling proved to be more difficult to export and were subsequently refused by Southeast Asian countries. When Southeast Asian countries refused to accept these Korean-made (plastic and vinyl) solid wastes, the damage of plastic waste went back to local residents living far from Seoul, the metropolitan area, and other large cities. Some plastic wastes were even left in the yard and/or the top of the house in the cities.

In front of a rural village in Uiseong-gun in Gyeongbuk province, not far from the upstream of Nakdong River, a garbage mountain that has the height of a 10th-floor apartment, can be noticed from afar. A recycling company dumped 172,000 tons of waste there, which was 80 times the volume that it was permitted to dump (2,157 tons). It meant to sort out more than 60,000 tons of garbage of which 26,000 tons of plastic waste was meant to be taken out in 2019. However, they left the waste unattended for two years starting from 2016 to 2017. This meant that the national government spent a tax of 5.3 billion Won (US$ 4.8 million) with the additional budget to dispose of about a third of the 170,000 tons of waste produced. In addition, 5 billion Won (US$ 4.5 million) was set for handling this mountain waste in Uiseong-gun (Lim, June 3, 2019). For expenses other than waste disposal, such as treatment expenses for residents, the county had to make appropriations for additional budget. Due to the disordered waste management system, more than 10 billion Won in taxes were only spent on the garbage disposal.

On the other hand, in a village in Cheonan, Chungcheongnam province, about 13,000 tons of wastes were left without intervention for over a year, thus causing nearby residents and administrative authorities to suffer as a result. It is estimated that it will cost about 4 billion Won to clean up the wastes, thereby fueling growing concerns about further wasting taxes (Han, Dec. 21, 2020).

Since it relies solely on the performance reports of solid waste management companies, it becomes difficult to recognize the flow of wastes into cities, especially when the waste disposal company's report are often invalid and cannot be trusted as they may hand over the waste disposal to other companies. This happens unless it is often monitored by related organizations, like the local governments. For example, although more than 170,000 tons of wastes were dumped in Uiseong, the amount of waste reported to the government was no more than 50,000 in 2017 (Lim, June 3, 2019).

Nonetheless, even if the local government finds out that the volume of waste being kept there exceeds the permitted volume, administrative disposition of notification

Source: This photo was taken by the author on October 2, 2020.
(Source mention the photo taken by the author)

Fig. 3. The Trash Mountain in Uiseong.

is difficult and the sanctions are weak, thus making it conducive for waste disposal companies to simply dump garbage and get away without any ramifications.

He further added:

> After the controversy over the garbage mountain, the county conducted special inspections and thorough investigation. But manpower was limited for each investigation. Since it was a rural village, I think it would not be noticeable if they throw something there, so they simply dump the trash there. In the end, it was not the responsibility of those who discarded the wastes anymore, but the residents and local governments of the abandoned area. Unfortunately, the cost of incineration of the trash had to be paid with the people's precious taxes (Lim, June 3, 2019).

To address the difficulties of holding anyone responsible for the illegal disposal of wastes, the government intends to hold everyone, including the consignee and the landowners, who are involved in the illegal disposal process in the future and punish them for such offenses, that is instead of simply fining to imprisoning them (Kim, August 9, 2019).

To curb the situation, waste disposal companies need to seek permission from the Ministry of Environment before changing their company's name so that person/ companies discharging illegal waste can be held responsible and penalised at once. If waste is disposed of through administrative execution, it is allowed to check the property first to prevent property concealment.

Since 2019, the number of national garbage mountains that the Ministry of Environment has identified totals 356 places, 1.15 million tons (as of the end of August, 2020). The ministry should be strict toward the illegal dumping of waste. However, the volume of garbage does not decrease. A new garbage mountain can be still seen day after day (Moon, Oct. 29, 2020).

Until recently, the government had discovered a total of 437 garbage mountains and 1.91 million tons nationwide. By August of 2022, 1.578 million tons of them were cleared. It has been confirmed that 54 (436,328 tons), half of the 108 garbage mountains nationwide that local governments cleaned up by proxy, were charged to innocent victims (landowners). The amount of compensation claimed amounts to 33.7 billion won (Lee, Dec. 9, 2022).

The problem of the saturation of landfills in the Seoul metropolitan area shows that Korean society is at a critical point so far as the solid waste management issue is concerned. This was because domestic wastes, particularly plastic wastes, have been rapidly increasing due to the rising number of single households. Korea's Ministry of Environment estimated that domestic wastes were 41,000 tons per day in 2017. This is 30,000 tons higher compared to the data five years ago.

According to the report of the European Plastics and Rubber Association (EUROMAP), Korea's annual plastic consumption per capita in 2015 was 132.7 kg, which was the highest in the world at that time. Belgium topped the list with 177.1 kg. Based on the plastic used for packaging, Korea is the second country (132.7 kg) after Belgium in 2015. Korea's Ministry of Environment estimated that the number of living plastic waste generated in the country in 2018 in 3.23 million tons, which was 71.7% of what it was 10 years before. During the same period, the total amount of household wastes also increased by about 10%. In particular, the increased rate, which stayed at 10.6% for five years until 2013, soared to 46.6% over the next five years. In 2020, plastic wastes are expected to increase by 14.6% from the previous year due to preventive measures that need to be taken against the corona virus infection (COVID-19) (Kim, Jan. 5, 2021).

The Big Trouble With Trash Due to the Ban on Importing Wastes into China

In April 2017, the initiative of circular development has been announced with the aim of strictly controlling the import of wastes and increasing the level of domestic waste recycling. After that, in January 2018, China stopped importing 24 kinds of wastes, including plastic wastes, recyclable paper, and so on. Up to this period, China was a major importer of plastic wastes, accounting for more than half of the world's waste imports. However, it can no longer ignore its own environmental pollution, thus stressing the importance of circulating resources.

As the demand for plastic wastes in China declined, the price of recyclable wastes plummeted. From 148 Won in April 2017, the price of recyclable paper fell to 90 Won in September 2017, or six months thereafter. This made collecting recycled wastes unprofitable for the waste collectors. Eventually, recycling companies halted waste collection across the country. Private collection companies indicated that they would no longer collect wastes because of losses (Hong, 2018). This began the so-called 'garbage catastrophe'.

Since 2017, China has continuously reinforced its import waste regulations by amending its Solid Waste Act and enacting relevant administrative regulations to protect its environment. On December 18, 2020, the Ministry of Environment in Korea announced that the Chinese government will ban the importation of all solid wastes from January 2021, in accordance with the new enforcement of its revised 'Solid Waste Environmental Pollution Prevention Act'. The Ministry of Environment estimated that the volume of solid wastes exported by China from Korea is 14,000 tons per year. This is 93% less than the volume that it used to import before it tightened its waste importation regulations in 2017. The export of household waste plastics to China has been suspended by Korea since 2018.

Table 1. Chronology of Trash Concerns in Korea.

January 2018	China stopped importing 24 kinds of waste, including plastic wastes
March 2018	Domestic recycling garbage prices fell
April 2018	Recycling companies stopped collecting solid wastes
After April 2018	Outbreak of big trouble with the proliferation of solid waste issues
March 2019	Vietnam banned imports of plastic wastes, effective from 2025
2019 Onwards	Ongoing trash crisis

"[Land of garbage] ④ How did we become 'garbage Korea'?," KBS News, April 3, 2019.

Currently, all solid waste export in Korea comprises wastes coming from business sites, such as slags and dust. Because these wastes are treated domestically or in third countries as part of the responsibility of the trash emitter, the Ministry of Environment predicts that the direct impact of this in the market will be economically insignificant. However, the domestic solid waste problem was already at saturation point, and China's importation ban on recyclable waste products has exacerbated the problem.

Trash Destination Gone; Where Will These Go?

With the export road to China being blocked, the government looked to Southeast Asia as an alternative destination for its trash. However, it is also strengthening environmental regulations as well. In July and October 2019, the illegal exportation of garbage to the Philippines was caught twice by a local environmental civil group, the Eco-Waste Coalition of the Philippines. Korea's prestige in the international market was significantly reduced. Vietnam, a major waste importer in Southeast Asia, later announced that it would completely ban the import of waste plastics from Korea by 2025 onwards.

Another factor that contributed to the Korean trash crisis was the adoption of the volume-based waste rate system in 1995. On January 1, 1995, the volume-based waste rate system was implemented along with a system for separate discharge, collection, and sorting of recyclables. However, this failed to effectively separate different types of plastics. In order to promote the separate disposal of recyclables while envisioning a volume-based waste rate system, the local governments decided to collect recyclables free of charge without paying garbage fees. Once the pay-as-you-go system was implemented, economical recyclable items such as paper wastes, scrap metal, and used clothes were bought by private recycling companies through direct contracts with the apartment residents. However, as China's imports of recycled products increased rapidly after 2000, the price of recycled products including plastics soared. Mixed plastics, which were not economical in the past, were also included in the items being bought by private recycling companies.

According to Article of the Waste Management Act in Korea, the legal responsibility for the collection and recycling of waste lies with local governments. However, Korea's recycling market is not active in supporting the state, but the private sector. It is being criticized that private businesses are passing on the cost of recycling, which is the responsibility of local governments, to the private sector, even if they cannot form a recycling market on their own.

As the producer responsibility recycling system took effect in 2004, vinyl wastes were added to the list of recyclable products that must be separated and identified for export. Since vinyl wastes were not paid products in the recycling market, local governments collected and separated them while the business sector such as producers supported their costs. However, due to fierce competition among private businesses for recycled products from apartment houses, a market structure was established wherein private businesses collected and separated these despite the abolition of collecting vinyl wastes. This has not changed due to the long-term recycling market boom in the 2000s (Hong, 2018: 2).

According to the revised Resource Recycling Act, the use of disposable vinyl has been banned in supermarkets starting from 2019. This is one of the follow-up measures adopted as part of the implementation of the comprehensive recycling waste management policy and the revision of the enforcement regulations of the Act on the Promotion of Saving and Recycling of Resources. This promoted the prohibition of the use of plastic bags in large-scale stores (about 2,000 locations) and supermarkets (about 10,000 locations) (Lee and Jeong, 2019).

Korea is going to incinerate plastic wastes and use it for a thermal power plant to extract petroleum. On the other hand, the European Union (EU) and others see only 'recycling materials' that are used by changing plastics into raw materials and turning them back into plastics. According to Greenpeace, although it is told that the recycling rate of Korean plastic waste was about 62.0% as of 2017, it was only 22.7% of the recycled materials. In the same year, the EU's recycling rate was about 40%.

Impact of COVID-19 on the Use of Plastics

With the spread of the coronavirus-19 (COVID-19) toward the end of January 2020, the Ministry of Environment temporarily allowed the use of disposable items in food service establishments, such as coffee shops and restaurants. Temporary exceptions were granted when the level of infectious disease crisis warning under the current law is at 'borderline' or higher. Since the infectious disease crisis warning is currently at the 'severe' stage, the use of disposable items is permitted. This caused a concern that the volume of solid wastes would again rise due to the increase in the use of disposable items. Indeed, in the first half of 2020, vinyl production increased by 11.1% when compared to 2019. Plastic production increased by 15.16%. Adding to this concern was the fact that the plastic recycling market is struggling due to the economic downturn and falling international oil prices (Lim, Sep. 15, 2020).

Due to COVID-19, the use of disposable products is recommended for personal hygiene and safety. The use of disposable products is likewise becoming a daily routine, raising concerns about environmental pollution.

Taking these into consideration, the Korean government allowed the use of disposable items in coffee shops and restaurants but were 'subject to restrictions' effective from January 2020. Because of the prolonged COVID-19 pandemic, however, it has led to an increase in plastic wastes. Packaging materials also have increased sharply since online shopping, which relies on courier services, has become popular. As the number of parcels and the number of delivered food packages increase, the volume of household wastes, such as plastic, vinyl, paper, and foamed resin, that use packaging containers likewise increases (Kim, Sep. 12, 2020).

According to the petrochemical industry, disposable plastics such as plastic gloves and door handle covers are also becoming essential safety supplies during the quarantine against COVID-19. The Ministry of Environment estimated that the use of plastic packaging materials during the first quarter of 2020 (January to March) increased by 20%. Paper wastes and plastic packaging also increased by 15% and 8%, respectively. For 'A-list' companies producing vinyl products, the volume of sales of vinyl gloves surged by more than 170% in April 2020 compared to their volume of sales during the same period in 2019 (Heo, July 22, 2020).

With the increased use of disposable products, since the COVID-19 pandemic, environmental experts warned that another 'garbage crisis' would be likely to occur in Korea. This necessitates the examination once again of the segregation methods being adopted in Korea. How can people properly segregate solid wastes? Why is it important to accurately to dispose of or discharge wastes properly?

As experts say, the reason why 'accurate' segregation and disposal is important is that even though recycling is not possible, randomly separated wastes not only interfere with the sorting process but also damage the machines used in the recycling process. Since December 2020, in particular, transparent PET bottles have been collected separately in large complexes by security guards and cleaners, who are already overburdened to do this work. These bottles have much higher recycling value than other colored plastic bottles, so these are collected separately and made into clothes or shoes.

First, to reduce the production and the use of plastic containers, it is recommended to set the production ratio of plastic containers among containers produced for container manufacturers over a certain size. To this end, in 2022, in accordance with the Framework Act on Resource Recycling, the recycling availability evaluation system was used to assess whether resource recycling is easy for each company. The production target for plastic containers that are relatively difficult to recycle is lowered, and glass bottles that are advantageous for reuse or recycling are produced to raise instead. The ratio of plastic containers among all containers is aimed to be reduced from the current 47% level to 38% by 2025 (Kwon, June 4, 2020).

There are also some attempts to reduce using plastics by the business sector. For instance, 'Market Cully', the online food market, which is said to be a pioneer of early morning delivery, has been pinpointed that it is causing environmental pollution due to excessive packing. Accordingly, the styrofoam boxes used for packaging the frozen products delivered by *Saebyul* were converted into paper. All the packaging materials for delivery products, such as cushioning packaging, box tape, and zipper bags, are also changed into paper, minimizing the use of plastic and Styrofoam; all packaging materials will be converted to paper materials by 2021 all packaging materials was converted to paper materials by 2021. Furthermore, no more than the carrier bags made by paper have been used since November 24, 2022.

Coastal Garbage Collection

According to the Korean coastal waste monitoring results, plastics account for 56% of all litter. By category, it is found that the ratio of PET bottles, plastic bags, caps, ropes, and steel foam buoys is high among all plastics (Lee, 2018). According to the KOIST, MPs were also detected in four types of shellfish, such as oysters and clams, that are sold in Korea (KIOST, 2017).

Marine waste collection can be largely divided into projects carried out by the central government and local governments; it depended on the nature of the marine waste purification project, offshore sedimentation waste collection project, fishing ground environment improvement project, oil-damaged area support project, illegal fishing nets demolition project, and so on. In 2017, the volume of garbage collected on Korean coasts was 82,175 tons. In 2019, it rose to 108,644 tons. It increased by 32.2% in two years. The share of the East Coast dropped sharply from 50.8% in 2017 to 42.8% in 2019 and then rose again to 49.6% in 2019, which was almost similar in terms of volume to the garbage collected on the West Coast. Moreover, the volume of plastic wastes around the southeast coast where the population was larger, was about twice as much as that of the east coast or the west coast region, where the population was less. About 86% of the plastics collected were microplastics (MPs) of less than 300 micrometers (μm). Most of the MPs found on the southern coast came mainly from styrofoam buoys (KIOST, 2015).

The volume of plastics flowing into the ocean from the *Nakdong* River is 53 tons per year, reaching about 1.2 trillion pieces. It is estimated that the damage costing over 30 billion won was due to plastic wastes coming from *Geoje* Island alone due to local heavy rains and typhoons in 2011. Marine pollution caused by plastics also led to a decrease in the number of tourists that go to the area (Shim, 2019).

Garbage Inflows and Outflows into the Korean Ocean

According to the data collected from 2007 to 2014, plastic wastes accounted for 37%, polystyrene foam, 35% of wood wastes, and 12% of floating marine litter (FML) in Korea. The main plastic trash composition includes plastic bags, bottles, plates, and ropes. Floating marine debris was found in tourist spots, ports, fishing areas, industrial, and recreation areas along the coast of China, which flows into the coast of Korea (NOWPAP MERRAC, 2017).

For the most part, plastic wastes and styrofoams were the highest in terms of the volume of debris. Plastic wastes made up almost (or more) half of the marine wastes. Wood, glass, tobacco, or foreign marine debris account for the rest of the debris (NOWPAP MERRAC, 2017: 30). About 20–30% of the total marine debris collected from 2010 to 2015, which varies slightly depending on the year, were fishery-related wastes such as styrofoam/plastic buoys or ropes. The rest (about 70–80%) of marine debris originated from land.

In general, the volume of marine debris tends to increase during the summer months. In Korea, geographical location, heavy rain, and summer season (June and July) generate more wastes in Northeast Asia. Marine garbage increases during typhoon season (August/September). The volume of marine debris collected is generally higher on the southern coast (i.e., Tongyeong, Masan, and Jeju). The aforementioned study noted the dominance of EPS particles in Geoje Island in the Southeast Sea (Heo et al., 2013). Other studies indicated a large portion of the MPs from the sea level suggesting that EPS buoys for aquaculture may be an important source of such (Song et al., 2015).

It has been confirmed that this trash floated south along the east coast along the ocean currents. The research team predicted and traced back the path of the movement of objects floating in the sea along the currents and winds with the floating material movement diffusion model and confirmed that the garbage found in the *Gangneung* coastal area on September 17 came from near the Tuman River around August 28, 2016 (Park & Kim, Nov. 4, 2016).

Conclusion

The rise of plastic in the environment compelled the Korean government to commit to cutting its plastic wastes in half by 2030. Unfortunately, due to COVID-19, the current governmental policy was not unprecedentedly stable, despite the possibility of increased impacts of disposal of masks and gloves on the aquatic and terrestrial ecosystem. Another trash crisis may come from the burden of small waste selection and recycling businesses due to the rise of minimum wage. Even so, according to the 2050 Declaration of Carbon Neutrality, the transition to a 'post-plastic society' remains in full swing. The government has established measures to reduce plastic generation over the entire cycle, such as reducing plastic use by 20% by 2025 and increasing the current recycling rate of waste plastic from 54% to 70%. At the same time, the importation of plastic waste from abroad is completely proscribed from 2022 (Lee, Dec. 24, 2020).

It is stressed that for Korea to dispel the stigma of being the world's worst plastic abuse nation, public awareness of reducing plastic use should be improved. The campaign to reduce the use of disposable plastic has been conducted at the national level since 2018. After all, if humans cannot live without discharging trash, they have to live with it. However, it has to first establish and enforce regulations that promote appropriate methods of waste segregation. Companies will change only when consumers strengthen their civic awareness and lifestyle through proper separate discharge education and speak out based on this fact.

In addition to the importance of segregating individual wastes, the advancement of corporate plastic recycling technology is of paramount importance. In fact, the lack of sharp countermeasures is even more problematic, and it must be solved for the next generation. Although measures such as the regulation of disposable products around the world were already at hand, these cannot be the fundamental solution to plastic pollution. To solve plastic contamination fundamentally, disposable plastics must be reduced from the production stage with a certain responsibility. In addition, to promote the use of recycled raw materials, the Ministry of Environment decided to reduce producer-responsible recycling contribution in a proportion to the amount of recycled raw materials used by producers. Urging improvement of the material of containers and packaging considering recycling at the production stage; practical regulation is needed. It would be mandatory for local governments to purchase more than a certain percentage of recycled products made from recycled raw materials. Lastly, the central government's coordination and integration of waste-related work between ministries, as well as institutional support for plastic recycling, is very crucial as this is most effective when combined with individual incentives.

Acknowledgment

This work was supported by National Research Foundation of Korea Grant funded by the Korean Government (NRF-2021S1A5A2A01068565).

References

Choi, Byeongseong, "Terrible oyster farm. It must be like this", *Ohmynews*, December 21, 2020.

EUROMAP. 2016. *Plastics Resin Production and Consumption in 63 Countries Worldwide.*

Han, Sol, "13,000 tons of waste are left in the mountains behind the village... "Only 4 billion won in processing costs", *KBS*, December 21, 2020.

Heo Dong-jun, "Disposable products that became essential for quarantine...US-Europe also postponed 'plastic regulation'", *Dong-A Ilbo*, July 22, 2020.

Heo, N.W., Hong, S.H., Han, G.M., Hong, S., Lee, J., Song, Y.K., Jang, M. and Shim,W.J. 2013. "Distribution of Small Plastic Debris in Cross-section and High Strandline on Heungnam beach, South Korea." *Ocean Sci. J.* 48: 225–233.

Hong, S., Lee, J. and Lee, J. 2020. *Understanding of Floating Marine Litter Sources and Flows in the NOWPAP Region.* NOWPAP MERRAC. Sinkwangsa.

Hong, Suyeol. 2018. "T-he 'Garbage Disruption', which became an opportunity for telephone upset," *World and City*, pp. 18–27. (in Korean)

Hyuck Jun Kwon, "'Deplastic society' in earnest… 20% reduction in plastics by 2025", *News1* December 24, 2020.

Kim, Hyeong Jong, "Korea's plastic usage is 'notorious', nowhere to bury," *Hankookilbo*, January 5, 2021.

Kim, Jeongyeon. "Get rid of all the nationwide garbage mountains this year... Strengthen punishment for illegal emitters." *Daily Joongang* August 6, 2019

Kim Joo-hyun, "Sushi was delivered, so full of plastic...the trash attack", *Money Today*, September 15, 2020.

KIOST. 2015. *A Study on Pollution of Coastal Environment by Microplastics*, Ansan: Korea Ocean Science Technology.

KIOST. 2017. *Research on Food Safety Management for Micro-Plastics.* (in Korean)

Korea Institute of Ocean Science and Technology. 2015. *A Study on Pollution of Coastal Environment by Microplastics*, Ansan: Korea Ocean Science Technology.

Kwon Jin-kyung, "The 'garbage mountain' that became a laugh...Rejected Korean plastic, why?", *KBS*, July 12, Collection of archival materials of the World Cup Park," Seoul Culture Today June 4, 2020.

Korean Ministry of Environment. 2020. *2019 National Waste Generation and Disposal Status.*

Lee, Chenga. "The disposal cost of 54 garbage mountains is 33.7 billion won.The innocent landowners took on their shoulders." Donga-ilbo. Dec. 9, 2022. "Limits and Implications of Waste Plastics Management Policy." Issue and Diagnosis, No.368, Korea Economic Research Institute. (in Korean)

Lee, Jeong-Yong, "Incheon City, Garbage Independence's 'Signal Bullet' Shot", *Sisa Journal*, No. 1625, December 5, 2020.

Lee, Jongmyung. 2018. "Marine waste status and countermeasures." Beach usage culture improvement discussion, Ministry of Oceans and Fisheries, Marine Environment Corporation. (in Korean)

Lee, Seok-hyung, "'Deplastic society' in earnest… 20% reduction in plastics by 2025", *News1*, December 24, 2020.

Lim, Jin-hee. "The story of Uiseong people groaning in the garbage mountain < 2019 New Plastic Report>", *CBS Nocut News*, June 3, 2019.

Lim So-Hyun, "[Corona trash trash ②] I ordered food for delivery… Disposable Bomb", *New Daily Economy*, September 15, 2020.

Marine Innovation, "Plastic Waste, it is OK?," https://100up.kakaoimpact.org/problems/49/view (Search: Jan, 20, 2021).

Moon, Dongseong. "Putting on a ganster and throwing it away…'national waste disposal cost of 100 billion won'" *Kookmin-ilbo* October 29, 2020.

Nam, Soo-hyun, "You say you can't separate disposable forks? As it is, a bigger crisis than the 'garbage catastrophe' two years ago", *JoongAng Ilbo*, November 11, 2020.

NOWPAP MERRAC. 2017. *Understanding of Floating Marine Litter Distribution in the NOWPAP Region*, Sinkwangsa.

Park, Young-min. "China, prohibits the entire import of solid waste. Increased possibility of domestic waste paper supply excess", *GDnet Korea*, December 18, 2020.

Park, Young-gyu and Kim, Gwang-seok, "Checking the route of movement to the Cheonryan Satellite". KIOST, Confirming the inflow of water waste from North Korea and China to the East Sea, Korea Institute of Ocean Science and Technology, November 4, 2016.

Pyo, Joo-Yeon. "[Attack on the continent ③] Marine debris from China, destroys the clean waters of the west and south seas," *Newsis*, March 30, 2016.

Shim, Won-jun. 2019. "Plastic environmental pollution." 2019 KRIBB Issue Conference, April 2.

Song, Y.K., Hong, S.H., Jang, M., Han, G.M. and Shim, W.J. 2015. "Occurrence andDistribution of Micro-plastics in the Sea Surface Microlayer." in Jinhae Bay, *South Korea. Arch. Environ. Contam. Toxicol.* 69: 279–287. <Newspaper and broadcasting articles in Korean> "[Land of garbage] ④ How did we become 'garbage Korea'?," *KBS News*, April 3, 2019.

The disposal cost of 54 garbage mountains is 33.7 billion won. The innocent landowners took on their shoulders." Donga-ilbo. Dec. 9, 2022.

Chapter 13

The Role of International Water Law in the Management of Marine Plastic Pollution

The Case of China and Its Transboundary Rivers

David J. Devlaeminck

INTRODUCTION

After establishing increasingly strict regulations regarding the import of plastics, China enacted a partial ban on the import of various kinds of plastic waste in 2018. As of January 1, 2021, China enacted a full ban on solid waste imports (Global Trade, 2021), and pledged to cut single-use plastics in the coming years (Xinhua, 2020). This ban sent shock waves through the global recycling industry, resulting in pile up of plastic waste in states around the world and even the halting of some local recycling programs (Semuels, 2019). Recyclables began piling up in plastic-exporting states as plastic importing states began to experience a rapid and drastic increase in imports of plastic waste (Agence France-Presse, 2019). While states around the world have struggled to cope, China's ban also acted as a catalyst, forcing plastic exporting states to rethink their plastic waste programs and enact their own policies regarding the import of plastics (Nguyen, 2020). While this has certainly shaken the industry, widespread use of disposable plastics remains the norm (at least for now). Since the development of plastics in the 1950s, the world has produced approximately 8.3 billion tonnes of plastic, producing around 300 million tonnes annually today (UNEP, 2020). Only 9% of this waste has been recycled, while 79% ends up in landfills, dumps, and the environment, with roughly 8 million tonnes ending up in the world's oceans every year (UNEP, 2020). As of 2015, it was estimated that there were approximately 150 million tonnes of plastic waste in the world's oceans. If no action is taken, this will

School of Law, Chongqing University, Chongqing, China.
Email: djdevlaeminck@cqu.edu.cn

rise to 600 million tonnes by 2040 (Parker, 2020). If the trend continues, there could be more plastic than fish in our oceans by 2050 (UNEP, 2020).

Although these plastics are in the world's oceans, a recent study indicates that 88–94% of marine plastics come from land-based sources. These plastics are carried by rivers into the ocean, and 90% of the ocean's plastic waste is carried by 10 rivers, including the Amur-Heilong (38,267 tonnes), Brahmaputra-Ganges-Meghna-Yaluzangbu (72,845 tonnes), Hai (91,858 tonnes), Indus (164,332 tonnes), Lancang-Mekong (33,431 tonnes), Niger (35,196 tonnes), Nile (84,792 tonnes), Pearl (52,958 tonnes), Yangtze (1,469,481 tonnes), and Yellow (124,249 tonnes) rivers (Schmidt et al., 2017; UNEP, 2020). Of these 10 river systems, eight are in Asia, including the Amur-Heilong, Brahmaputra-Ganges-Meghna-Yaluzangbu, Hai, Indus, Pearl, Lancang-Mekong, Yangtze, and the Yellow River. A common riparian state amongst all of these rivers is China. Whereas the Hai, Pearl, Yangtze, and the Yellow Rivers are domestic rivers fully within China's territory and therefore are not governed by international law; the Amur-Heilong, Brahmaputra-Ganges-Meghna-Yaluzangbu, Indus, and Lancang-Mekong are transboundary, shared between China and 11 riparian states (Bangladesh, Cambodia, India, Laos, Mongolia, Myanmar, Nepal, Pakistan, Russia, Thailand, and Vietnam). Although China's domestic rivers are a large source of marine plastics themselves, this is not to say that China is a major contributor to plastic pollution in its transboundary rivers. Instead, this chapter will argue that as a growing power, regional leader and common riparian, China has the potential to play a leading role in promoting regional and global cooperation to combat marine plastics. Given the nature of the world's oceans beyond state jurisdiction as a global commons and that a major source of marine plastics is rivers, resolving this issue will require significant global and regional cooperation and coordination across all levels, including state-state cooperation.

While there are certainly a variety of technical solutions available to us (Nikiema et al., 2020), international law can provide a framework for state-state cooperation and coordination in regard to global and regional problems. Although international lawyers highlighted the issue of marine plastic pollution in the 1990s (Nollkaemper, 1993), it has only received greater scholarly attention over the past few years.[1] Although efforts are underway to establish a binding global plastics treaty (UNEP, 2022),[2] international law governing marine plastics, like international environmental law generally, is significantly fragmented. However, in regard to marine plastics entering the oceans from rivers, there are two areas of international law that are of particular relevance, the law of the sea codified through the United Nations Convention on the Law of the Sea (LOSC,1982a), and the law of international watercourses, codified in the two global water conventions—UNECE Convention on the Protection and Use of Transboundary Watercourses and International Lakes

[1] For example, a special issue on the subject of "Plastics Regulation" was published in the Review of European Comparative and International Environmental Law, 27(3) (2018).

[2] In March 2022, states adopted a UNEP resolution where they committed to working towards an international plastics treaty. The draft is expected to be complete by 2025 (UNEP, 2022). Although the draft may be complete at that time, it is yet to be seen how quickly the treaty can achieve the required number of ratifications and enter into force.

(Water Convention, 1992a) and the United Nations Convention on the Law of the Non-navigational Uses of International Watercourses (Watercourses Convention, 1997a).

This chapter seeks to explore the role of international law in the governance of transboundary rivers as a source of marine plastics with an emphasis on the law of the sea, the law of international watercourses, and China's governance of its transboundary rivers. China and almost all of its riparian neighbours are party to the LOSC (1982a) and therefore bound by its provisions; however, of all the states on the four aforementioned rivers, only Russia and Vietnam are party to either of the global water conventions. While the global water conventions have limited state parties in Asia, China has established a series of binding bilateral agreements and various non-binding instruments with other nations through which it seeks to jointly govern its transboundary rivers. Although these agreements do not directly mention marine plastics or the marine environment, they may offer basic frameworks that can be further elaborated upon and progressively developed by the parties. Thus, the chapter will first unpack the role of international law in managing rivers as a source of marine plastics through the law of the sea and the law of international watercourses. It will then explore regimes related to the joint governance of China's transboundary rivers, highlighting the role of current legal regimes in the region and (often large) gaps in their application to marine plastics. A focus on China's treaties does not provide a complete picture of the governance mechanisms in place on all of these rivers, however it can highlight opportunities for China to take a leading role in the fight against marine plastics.

The Regulation of Rivers as Sources of Marine Plastics Under International Law

The international legal regime related to the protection of the environment has been notably fragmented, consisting of a series of overlapping sub-regimes at the global and regional levels (Koivurova, 2014). In regards to rivers as a source of marine plastics, there are a series of rules across sub-regimes of international environmental law that may apply, most notably the law of the sea and the law of international watercourses. Although they may apply to the issue of marine plastics, provisions are often vague in their application to rivers. The LOSC (1982a) sets out a series of obligations across a wide range of issues, including territorial delimitation, transit, economic rights, resource development, and environmental protection among many others. Part XII, Protection and Preservation of the Marine Environment, includes a series of obligations related to the protection of the marine environment. Article 194 requires all states to "take, individually or jointly as appropriate, all measures consistent with this Convention that are necessary to prevent, reduce and control pollution of the marine environment from any source, using for this purpose the best practicable means at their disposal and in accordance with their capabilities [...]" (LOSC, 1982a).

While plastic pollution is not specifically mentioned, the provision indicates that these measures "shall deal with *all sources of pollution* of the marine environment" and "shall include, *inter alia* [...]";[3] measures meant to minimize "the release of toxic, harmful or noxious substances, especially those which are persistent, *from land-based sources,* from or through the atmosphere or by dumping" (LOSC, 1982a, Art. 194(1)). Section 5, International Rules and National Legislation to Prevent, Reduce and Control Pollution of the Marine Environment, even goes on to include select provisions that address this issue directly (LOSC, 1982a, Sect. 5). In regards to land-based sources of pollution, Article 207 obligates states to take a series of measures to prevent, reduce and control land-based sources of pollution and "adopt laws and regulations [...] taking into account internationally agreed rules, standards and recommended practices and procedures" (LOSC, 1982a, Art. 207(1)), "take other measures as may be necessary" (LOSC, 1982a, Art. 207(2)), to "endeavour to harmonize their policies in this connection at the appropriate regional level" (LOSC, 1982a Art. 207(3)), and "endeavour to establish global and regional rules, standards and recommended practices and procedures [...] taking into account characteristics of regional features, the economic capacity of developing States and their need for economic development" (LOSC, 1982a, Art. 207(4)). Article 213 further obliges states to "enforce their laws and regulations adopted in accordance with article 207" and to adopt rules and regulations so as to implement other relevant international rules "through competent international organizations or diplomatic conferences" (LOSC, 1982a, Art. 213). While this broad approach recognizes the "complexity of [land-based sources of pollution]" and the "range and diversity of legal approaches required to address it", it is also admittedly vague and unclear (McIntyre, 2020).

Another area of international environmental law relevant to marine plastics from rivers is the law of international watercourses. This area of international law seeks to provide "an identifiable corpus of rules of treaty and customary law that determine the legality of State actions with respect to water resources that cross national boundaries" (Wouters, 2013), and has been codified and progressively developed in the Water Convention (1992) and Watercourses Convention (1997). Whereas the scope of the Watercourses Convention does not explicitly mention the connection between rivers and marine ecosystems in its definition of an international watercourse, the preamble of the Water Convention notes that rivers can act as a source of pollution for the marine environment (Water Convention, 1992a, preamble; Finska and Howden, 2018).These conventions are founded upon two substantive rules, i.e., the principle of equitable and reasonable utilization and the due diligence obligation not to cause significant harm. While they are the centrepiece of international water law, these rules remain largely unhelpful for the management of land-based sources of pollution. As codified in the Watercourses Convention, states are to "utilize an international watercourse in an equitable and reasonable manner" and "with a view to attaining optimal and sustainable utilization thereof and benefits therefrom" (Watercourses Convention, Watercourses Convention, 1997a, Art. 6). The Water Convention more strongly emphasizes pollution prevention, and it too includes the principle of equitable and reasonable utilization, obligating all states to "ensure that transboundary waters are used in a reasonable and equitable way, taking into particular account their transboundary character, in the case of activities which cause or are likely to cause transboundary impact" (Water Convention, 1992a, Art. 2(2c)).

[3] The phrase *inter alia* means among others.

In achieving equitable and reasonable utilization, states are to take into consideration as a whole a series of factors (Watercourses Convention, 1997a; Tanzi, 2015).[4] While these factors include "the effects of the use or uses of the watercourses in one watercourse State on other watercourse States" (Watercourses Convention, 1997a, Art. 6), the provision is largely unhelpful as it is targeted at specific utilizations of shared water resources and their potential effects, not pollution in general. Furthermore, the balancing of interests that it requires does not include the protection of the marine environment, nor does it take into consideration the interests of coastal states, even if those states are riparian states (Nollkaemper, 1993). The due diligence obligation not to cause significant harm requires that all "watercourses States shall, in utilizing an international watercourse in their territories, take all appropriate measures to prevent the causing of *significant harm* to other watercourse States" (Watercourses Convention, 1997a, Art. 7). Although plastic pollution in rivers may be recognized as "harm", for it to fall under Article 7 the harm in question must be considered "significant". This entails that the harm must be "capable of being established by objective evidence," and that there must be "a real impairment of use", including a "detrimental impact of some consequence upon, for example, public health, industry, property, agriculture or the environment in the affected State" (ILC, 1988, page 36). While marine plastics have been shown to impact marine life (Gall and Thompson, 2015), this impact falls beyond the scope of international water law as it occurs in the marine environment. Recent research has begun to dive into the effects of plastic pollution on riverine ecosystems, illustrating its toxic effects, hazards for marine life, and even harm to livelihoods (Emmerik and Schwarz, 2019). However, it will be difficult to argue that this falls under the due diligence obligation not to cause significant harm as it may be impossible to point to a specific utilization of the watercourse that is causing said plastic pollution. Some forms of harm, however, can be recognized as "per se unreasonable", and these may include those that endanger human health or those that cause irreparable and long-lasting harm (McCaffrey, 2019).

Given the lack of clarity in applying the two primary rules of the law of international watercourses to the issue of marine plastics, it is best to look to other provisions for potential legal avenues. Article 23 of the Watercourses Convention, *Protection and preservation of the marine environment*, requires States to "take all measures with respect to an international watercourse that are necessary to protect and preserve the marine environment, including estuaries, taking into account generally accepted international rules and standards" (Watercourses Convention, 1997a). Whereas the aforementioned due diligence obligation not to cause significant harm is limited to the prevention of harm to other watercourse States, Article 23, based on the common interest of states, is not limited as such and provides a "duty to manage the watercourse in a manner that does not harm" the marine environment (Finska and Howden, 2018, page 248; also see McIntyre, 2020). The Water Convention also includes provisions that directly relate to the protection of the marine environment,

[4] This list of factors to take into consideration is found in Article 6, Watercourses Convention. The Water Convention does not contain such a list; however, interpretative guides of the Water Convention point to the list of factors found in the Watercourses Convention (UNECE, 2013).

such as Article 2(2a) which obligates states to "prevent, control and reduce pollution of waters". This could be applied to coastal states, as Article 9(4) clearly provides riparian states with an obligation to cooperate with coastal states so as to reduce transboundary impact (Water Convention, 1992a; Finska and Howden, 2018).

Beside these binding agreements, there are a series of soft law instruments that can guide cooperation among states. Soft law includes non-binding instruments that, although they do contain commitments made by States, are non-binding and not readily enforceable via dispute settlement mechanisms (Boyle, 1999; Guzman and Meyer, 2010; Thürer, 2009). Such non-binding instruments have played a significant role in the development of international environmental law, have often acted as a precursor to more binding agreements and serve as guidance for current practice (Dupuy, 1990). In regards to rivers as a source of marine plastics, the *Global Programme of Action for the Protection of the Marine Environment from Land-Based Activities* (1995) is most notable. Aiming to "prevent the degradation of the marine environment from land-based activities by facilitating the realization of the duty of States to preserve and protect the marine environment", the Plan of Action sets out a series of national, regional, and international recommendations to "establish and strengthen voluntary multi-stakeholder action such key sectoral issues", such as radioactive substances, persistent organic pollutants, heavy metals, and litter, among others (McIntyre, 2020, page 285). Furthermore, it clearly recognizes the interconnectivity between rivers and the marine environments and, in relation to litter, their role in transporting plastic pollution to marine and coastal environments (UNEP, 1995). This connection between marine plastics and land-based sources has been echoed in more recent soft law instruments and related initiatives at the United Nations, such as the Sustainable Development Goals, a set of 17 goals and 169 targets for global development that all states will jointly and individually work towards (UNGA, 2016). Through Goal Number 14, all states have pledged to "conserve and sustainably use the oceans, seas and marine resources for sustainable development". Through Target 14.1, states pledge that "by 2025, [they would] prevent and significantly reduce marine pollution of all kinds, in particular from *land-based activities, including marine debris* and nutrient pollution" (UNGA, 2016). States have also pledged to follow up, review, and report the progress in regards to the SDGs (para. 47) and indicators have been established with Target 14.1 progress being determined by "plastic debris density" monitoring parameters which use earth observation and modelling data, national data from all states, and other additional indicators. While the parameters are related to plastics in the marine environment, "river litter" is also included (UNEP, 2021).[5]

Marine Plastics in China's Transboundary Water Agreements and Instruments

While the aforementioned global conventions can enable and facilitate cooperation on marine plastics to a degree, they have limited parties on the four rivers that

[5] It should be noted that "river litter" as a monitoring parameter is not clearly defined. As an "additional indicator", the Manual does not elaborate on it further (UNEP, 2021).

are studied here (see Table 1). Eleven of the 12 aforementioned riparian states (Bangladesh, China, India, Laos, Mongolia, Myanmar, Nepal, Pakistan, Russia, Thailand, and Vietnam) are party to the LOSC.[6] However, only Russia is party to the Water Convention and only Vietnam is party to the Watercourses Convention. While all remain bound by the substantive rules of equitable and reasonable utilization and the due diligence obligation not to cause significant harm as customary international law, non-party China and its neighbours are not bound by these agreements. Furthermore, even those states that are bound (Russia and Vietnam), are only bound in their relations with other parties to the conventions (VCLT, 1969, Art. 34).[7] Thus, their impact is limited until other states in the region acede. There are, however, a series of binding agreements and non-binding instruments through which China and its neighbours jointly manage their transboundary water resources, which may assist in managing sources of plastic pollution from these rivers. While at present there are no agreements that involve China on the Indus, there are binding bilateral agreements with states on China's northeast, i.e., Amur-Heilong River which is shared among China, Mongolia, and Russia; there are also various soft law arrangements with the states on China's southwest, especially on the Brahmaputra-Ganges-Meghna-Yaluzangbu,which is shared among Bangladesh, China, India, and Nepal, and the Lancang-Mekong that is shared among Cambodia, China, Laos, Myanmar, Thailand, and Vietnam (see Table 2). Treaties are to be interpreted "in good faith in accordance

Table 1. State Parties to the LOSC, Water Convention and Watercourses Convention Among the Riparian States Studied Here (Water Convention, 1992b; Watercourses Convention, 1997b; LOSC, 1982b).

	Convention on the Law of the Sea (1982)	**Water Convention (1992)**	**Watercourses Convention (1997)**
Bangladesh	✓		
Cambodia	*		
China	✓		
India	✓		
Laos	✓		
Mongolia	✓		
Myanmar	✓		
Nepal	✓		
Pakistan	✓		
Russia	✓	✓	
Thailand	✓		
Vietnam	✓		✓

*Cambodia signed but has not yet ratified the LOSC. It is therefore not obligated by the provisions of the Convention but shall not frustrate its purpose (VCLT, 1969).

[6] It is interesting to note that Laos, Mongolia, and Nepal are parties to the LOSC, even though they are landlocked states. This makes the obligations of LOSC (Section 5) particularly relevant for them.

[7] For an exploration of this issue in the context of international water law, see Spijkers and Devlaeminck (2021).

Table 2. A Selection of Relevant Binding and Non-Binding Instruments on the Rivers Studied Here.

River	Basin States	Document
Amur-Heilong	China, Mongolia, and Russia	1994 China-Mongolia Agreement on Protection and Utilization of Boundary Waters 1995 Agreement on the Establishment of the Consultative Commission for the Development of the Tumen River Economic Development Area and Northeast Asia 1995 Memorandum of Understanding on Environmental Principles Governing the Tumen River Economic Development Area and Northeast Asia 2001 China-Russia Treaty on Good Neighbourliness 2008 China-Russia Agreement on Management and Protection of Transboundary Waters
Brahmaputra-Ganges-Meghna-Yaluzangbu	Bangladesh, China, India, and Nepal	1993 China-India Agreement on Environmental Cooperation 2006 China-India Joint Declaration
Indus	China, India, andPakistan	
Lancang-Mekong	Cambodia, China, Laos, Myanmar, Thailand, and Vietnam	1995 Agreement on the Cooperation for the Sustainable Development of the Mekong River Basin 2016 Sanya Declaration of the First Lancang-Mekong Cooperation (LMC) Leaders' Meeting 2018 Phnom Penh Declaration of the Second Lancang-Mekong Cooperation (LMC) Leaders' Meeting 2020 Vientiane Declaration of the Third Mekong-Lancang Cooperation (MLC) Leader's Meeting

with the ordinary meaning to be given to the terms of the treaty in their context and in light of its object and purpose" (VCLT, 1969, Art. 31(1)). These agreements are admittedly vague and do not include provisions that directly address plastic pollution. However, the seriousness of the issue and the fact that these rivers are major sources of marine plastics provide an opportunity for the progressive development of these agreements via subsequent agreements, practices, or relevant rules of international law (VCLT, 1969, Art. 31(3)) and even soft law instruments, which can assist in solidifying consensus on best practice (Shelton, 2009). Adopting such an approach would provide China with an opportunity to update and align these agreements with existing instruments and practices, such as that of the LOSC, the global water conventions, the Global Programme of Action, and SDGs, among others.

The Amur-Heilong river: Binding agreements

Although China is not a party to either of the global water conventions, its transboundary water treaty practice incorporates both equitable and reasonable use and the due diligence obligation not to cause significant harm (Wouters and

Chen, 2013) among its other provisions. This is most apparent in the Amur-Heilong river, which is shared among China, Mongolia, and Russia where multiple binding agreements are in force. As earlier discussed, these substantive rules may be unhelpful in the management of marine plastics; however, these same rules in China's transboundary water agreements offer a broad and often unspecified scope which may cover riverine plastic pollution. Article 2 of the *China-Mongolia Agreement on the Protection and Utilization of Transboundary Waters* (1994) provides that "for the purpose of protection and equitable and rational use of transboundary waters", the parties shall cooperate in investigating and surveying boundary water quality, monitoring, and reduction of pollution among other actions (China-Mongolia, 1994). The *China-Russia Treaty on Good Neighbourliness* (2001) also incorporates equitable and reasonable utilization, indicating that China and Russia "shall carry out cooperation in the protection and improvement of the environment, prevention of cross-border pollution, the fair and rational use of water resources along the border areas" among other areas of cooperation (China-Russia, 2001). Whereas some agreements, such as the *China-Russia Water Agreement* (2008), limit this obligation to the threshold of "significant harm" (Vinogradov and Wouters, 2013), others take a broader approach. The *China-Mongolia Agreement* (1994), for example, requires that the parties "should jointly protect the ecological system of the transboundary waters and develop and utilize transboundary waters in a way that should not be detrimental to the other side" (China-Mongolia, 1994). While these provisions do not explicitly mention plastic pollution or the marine environment, given their broad and undefined scope they could facilitate cooperation on riverine and marine plastics. Marine plastics do not necessarily fall outside of their scope, and they could assist in limiting plastic pollution if the parties were to interpret them to do so. Terms such as *detrimental to the other side, pollution, protection and improvement of the environment*, and *transboundary impact*, for example, require further development and exploration by the parties.

Apart from the substantive rules, there are also a series of procedurally oriented environmental obligations that could guide action on plastic pollution. For example, the *China-Russia Agreement* (2008) indicates that in order to protect and use the shared freshwater resources, the parties should cooperate in a series of issue areas [Art. 2(1)]. In regards to pollution, the parties are to conduct joint research and identify sources of pollution and take measures to control and reduce their transboundary impact [Art. 2(16)] (Vinogradov and Wouters, 2013). The *China-Mongolia Agreement* (1994) also provides for a series of obligations, including monitoring of pollution and water quality levels (Art. 2(4) and 3(1)). It also obligates the parties to "take measures to prevent, mitigate and eliminate the possible damage to the quality, resources and natural dynamics of the transboundary waters and aquatic animals and plants caused by natural or human factors" (China-Mongolia, 1994, Art. 6). While the provision lists "flood, ice run and industrial accidents", as examples, it is not limited to these instances (China-Mongolia, 1994). Both agreements establish a joint body, the China-Mongolia Joint Committee on Transboundary Waters (China-Mongolia, 1994) and the China-Russia Joint Working Group (Vinogradov and Wouters, 2013), which could act as a focal point of discussion on marine plastics, assist in implementation, and work to progressively develop these agreements.

Besides these transboundary water agreements, there are other cooperative institutions that have been established via a legal agreement that encompasses the shared watercourse. The Greater Tumen Initiative, first established in 1991 under the United Nations Development Program as the Tumen River Area Development Program, seeks to establish "their common interests" and "increase mutual benefits, to strengthen economic and technical cooperation, and to attain greater growth and sustainable development" for the states involved. It later evolved into the Greater Tumen Initiative in 2005 (Lee et al., 2019). While the Tumen River is only shared among North Korea, China, and Russia, the Greater Tumen Initiative includes Mongolia and South Korea as part of the region in which the parties will cooperate and enact projects and programmes for joint development (GTI, 1995a). The *Agreement on the Establishment of the Consultative Commission for the Development of the Tumen River Economic Development Area and Northeast Asia* (1995) indicates that parties "shall implement the Agreement on the basis of the international law governing relations between States" (GTI, 1995a), highlighting the importance of relevant international law, such as the LOSC, the Water Convention which Russia is a party to, and various other bilateral agreements. It also provides for the establishment of a commission (GTI, 1995a). In order to "attain growth and sustainable development" (GTI, 1995a), the member States have established the *Memorandum of Understanding on Environmental Principles* (1995). Although it is non-binding, states have pledged to jointly establish "environmental mitigation and management plan[s]" (GTI, 1995b). These plans are to include measures for the protection of:

Land resources, particularly wetlands, *fragile coastal areas*, forests and sensitive ecosystems; *preservation of biodiversity*, including threatened or endangered species and their habitats; [...] protection and improvement of *water quality*; protection of the *marine environment and marine living resources;* sound *disposal, management, treatment and movement of hazardous and solid wastes*; [...] and *monitoring of pollution* and environmental conditions (GTI, 1995b).

Similar to the Tumen Agreement, it also reaffirms the importance of international law and various multilateral agreements; although none are specifically mentioned (GTI, 1995b). Furthermore, the parties meet annually and release a declaration (GTI, 2021), providing an opportunity to refine and perhaps redirect their cooperation to emerging issues.

The Indus, Brahmaputra-Ganges-Meghna-Yaluzangbu and the Lancang-Mekong: Non-binding instruments

While China has agreements in its northeast, elsewhere cooperation is not facilitated through legally binding agreements. Although China does not have any agreements or instruments regarding the Indus River shared with India and Pakistan, China has established a series of soft law instruments that guide its cooperation on the Brahmaputra-Ganges-Meghna-Yaluzangbu and the Lancang-Mekong. In the Brahmaputra-Ganges-Meghna-Yaluzangbu, China has signed various soft law declarations with India. These instruments, however, are incredibly broad in scope covering nearly every area of cooperation between the two states. The *China-India*

Agreement on Environmental Cooperation (1994), however, includes pledges to cooperate on global environmental issues, including pollution control and "any other areas that may be agreed upon by the parties" (China-India, 1993). As this cooperation has unfolded, the two states have signed a series of declarations on various issues, some of which reaffirm their commitment to cooperation. The *China-India Joint Declaration* (2006), for example, provides that these two states will have regular consultations on various issues, including environmental degradation (China-India, 2006). There are, however, no relevant agreements with Bangladesh or Nepal. Whereas soft law cooperation with China and its Brahmaputra-Ganges-Meghna-Yaluzangbu riparians is limited, China has taken the lead in cooperation in the Lancang-Mekong. The four lower riparians of the Lancang-Mekong (Cambodia, Laos, Thailand, and Vietnam) are parties to the *Agreement on the Cooperation for the Sustainable Development of the Mekong River Basin* (1995) which established the Mekong River commission (MRC) (Mekong Agreement, 1995). China and Myanmar, however, are not parties. While the MRC has taken notice of the plastic pollution issue in the Lancang-Mekong by initiating a review of the issue of plastic pollution and regional legal and institutional frameworks as part of a project with the United Nations Environment Programme (MRC, 2020), China is not bound by the Agreement and is unlikely to play a major role in this process. Although China remains bound by customary law on these rivers, it also has sought to establish a series of soft law instruments, most notably the Lancang-Mekong Cooperation Mechanism (LMC).

The LMC, established by China in 2015,[8] is a multilateral institution counting all Lancang-Mekong riparian states as members (Cambodia, China, Laos, Myanmar, Thailand, and Vietnam). While it is a non-binding regime which has been developed via a series of soft law instruments including declarations and joint communiqués, its broad scope and complete coverage of the basin can act to catalyze action on the issue of marine plastics. Similar to the Greater Tumen Initiative, the LMC recognizes and operates alongside the international legal obligations of its member states, as LMC documents consistently exhort "respect for the UN Charter and international laws" (LMC, 2016; 2018; 2020), recognizing existing legal frameworks in the region. The LMC offers a broad scope via its 3 + 5 approach, which works across three pillars: (i) political security, (ii) economy and sustainable development, and (iii) social and people-to-people and cultural exchange. There are also five priority areas: (i) connectivity, (ii) production capacity, (iii) cross-border economic cooperation, (iv) water resources, and (v) agriculture and poverty reduction (MFA, 2016).[9] Such a broad scope may allow for a multi-pronged approach to combatting plastic pollution, offering a path for the reduction of plastic use, managing land-based sources of plastic, and mitigating plastic in the riverine environment. Even though LMC documents contain a strong emphasis on sustainability and environmentally conscious development, there is no mention of plastics or the marine environment.

[8] There are multiple dates that could be used as a starting point for the LMC. The LMC was first proposed in 2014, its first Foreign Minister's Meeting was held in 2015, and the First Leader's Meeting was held in 2016 (Devlaeminck, 2021).

[9] For a complete discussion of the approach of the LMC, see Devlaeminck (2022).

Sustainable development is a common theme throughout LMC documents, including the *Sanya Declaration of the First Leader's Meeting* and *Phnom Penh Declaration of the Second Leader's Meeting,* both indicating that LMC states will take action to "promote green and sustainable development" (LMC 2016; 2018). While there is no mention of plastic pollution of the marine environment in any LMC water-related document as of the end of 2021, given the *softness* of the mechanism it is certainly possible that it could develop in such a direction. The LMC provides a series of institutional arrangements in which it could do so, including bi-annual leaders meetings, annual foreign minister's meetings, and working group meetings (MFA, 2016); each of which can act as a forum of discussion for the management of marine plastics on the Lancang-Mekong and reflect this in future declarations or communiqués. Furthermore, given the growing relationship between the MRC and LMC (for example, MRC-LMC, 2019), this may be an area of future collaboration while also ensuring that action on marine plastics takes place at the basin level.

Conclusion: Legal Avenues for China in the Fight Against Marine Plastics

China is an incredibly important country in terms of Asia's transboundary rivers and a common riparian on nearly all of the major rivers that transport plastics from land-based sources into the marine environment. However, given the remoteness of these transboundary rivers in its territory, it is not necessarily the case that China is a major contributor to marine plastics in these rivers. That being said, given China's upstream position, its role as a regional (and increasingly global) power and its increasing engagement with international law, it can certainly play a larger role in combatting marine plastic pollution within its own borders and in coordination with its neighbours. China has illustrated that it is willing to take a leading role in international environmental issues, especially in environmental concerns such as climate change (Shen and Xie, 2019). Assuming a leadership role on the issue of marine plastics can provide a focal point for cooperation on China's transboundary water resources while aligning its transboundary practice with its own domestic actions to control plastic pollution. It could also offer an opportunity for China to actualize its concept of "ecological community", which seeks to transcend Western models of industrialization and establish a society that aspires to place sustainable development and co-existence with ecosystems at its core (Barresi, 2020; Corne and Zhu, 2020). In order to fully realize this, however, China will have to work closely with the international community and its riparian neighbours to close gaps in management both at the international and regional levels.

At the global level, there is significant fragmentation in conventions that are meant to enable cooperation on marine plastics. While the LOSC has 168 parties (LOSC, 1982b) establishing near-global coverage, the two global water conventions have significantly fewer parties. At the time of writing, the Water Convention, now open globally for accession, has 47 parties (Watercourses Convention, 1997b), whereas the Watercourses Convention has 37 parties (Water Convention, 1997b). While almost all of the states on the four rivers studied here are party to the LOSC, only Russia and Vietnam are party to the global water conventions.

How might we overcome this gap? Further uptake of these legal norms is required for them to act as adequate frameworks for the reduction of marine plastics in the region and elsewhere. Some have suggested the establishment of a global plastics treaty, which would establish norms instead of relying on the fragmented nature and partial coverage of current international law applied to marine plastics (Keselica, 2020; Kirk and Popattanachai, 2018). In 2022, the international community took a major step in this direction when states adopted a resolution at UNEP agreeing to initiate the drafting process of a binding convention on plastics, with completion of the draft by 2025 (UNEP, 2022). Such a treaty will "include both binding and voluntary approaches, based on a comprehensive approach that addresses the full lifecycle of plastic" (UNEP, 2022). At the time of writing the treaty appears to take the approach of the Paris Agreement, establishing basic objectives and allowing states to establish their own prevention, reduction, and elimination strategies (Kirk, 2022).

Regardless of its approach, this process could offer an opportunity to bridge fragmented international legal regimes related to marine plastics, including the Law of the Sea and the Law of International Watercourses. However, given that the large majority of marine plastics comes from land-based sources transported to marine environments by rivers, the discussions surrounding the drafting of a global plastics treaty are certain to overlap with discussions surrounding the norms of international water law. While both global water conventions are in force, they are not necessarily the most popular as mentioned earlier. China, a common riparian on rivers that are some of the largest sources of plastic pollution, was one of three states that voted against the adoption of the Watercourses Convention at the United Nations General Assembly in 1997 (UNGA, 1997). States, however, continue to be reluctant to adopt the package of norms provided by these conventions. This is evidenced by the 44 years it took in the study, drafting, and ratification before it entered into force, as the Watercourses Convention was first suggested for study by the International Law Commission in 1970 (UNGA, 1970) with a draft being voted on at the United Nations in 1997 (UNGA, 1997) and further 17 years before it entered into force in 2014 (Watercourses Convention, 1997b). Given that rivers are a major source of marine plastics and that there will be significant overlap between the norms of international water law and a marine plastics treaty, the process of study, drafting, and entry into force required for any such treaty may require a similar length of time. Thus, while a binding convention is something that should be considered in the long term, in the short term we must continue to utilize this fragmented legal landscape.

Although China voted against the Watercourses Convention at the United Nations, this does not mean that it opposes key norms of the Convention as China prefers to govern its transboundary waters via bilateral agreements which embody them (Wouters and Chen, 2013). For the joint management of these rivers, China has established a series of bilateral agreements with its neighbours in the northeast. While, as illustrated, these agreements are broad in scope and leave much to be defined and progressively developed by the parties, various aspects of them can be utilized to assist in reducing marine plastics, particularly provisions regarding monitoring, pollution reduction, and joint research to identify sources of pollution. However, given the vagueness of the terms used in provisions of these agreements, such as equitable and reasonable utilization and the due diligence obligation not to cause significant harm, at least one of the parties would need to first raise the issue under the arrangement and seek agreement on its interpretation in light of their commitments under other agreements and customary international law. Given its domestic push to combat plastic pollution, China could link its transboundary and domestic practice and raise this issue with its bilateral treaty partners through the institutional bodies available to them. While this may not be within the scope of the treaties discussed here, soft law can play a role in overcoming this gap.

Soft law has played an increasingly important role in the management of China's transboundary waters and has particular benefits over binding legal regimes. While binding regimes can take significant time to draft and enter into force, soft law can be drafted and signed quickly as states are more willing to do so given its non-binding nature. Furthermore, after signature, it is more malleable and therefore more easily and quickly amended (Boyle, 1999; Guzman and Meyer, 2010; Shelton, 2009; Thürer, 2009). With one soft law mechanism in place, the LMC and one hybrid regime, the Greater Tumen Initiative, soft law could act as a mechanism for quick action regarding plastic pollution as both contain various environmental norms, procedural aspects, and could be quickly amended. Although it is non-binding, this does not necessarily mean that states would not abide by their commitments, as states are compelled by both reciprocity and reputation (Guzman and Meyer, 2010). A soft law declaration in the Greater Tumen Initiative or one of the transboundary water commissions on the Amur could provide greater clarity on legal cooperation in China's northeast. The same could be done in China's southwest if the LMC States were to prioritize marine plastics as an issue of cooperation. Given the broad scope of the LMC, it seems an obvious fit and may even act as a pathway for cooperation between the MRC and LMC. On the Indus and Brahmaputra-Ganges-Meghna-Yaluzangbu, however, there are no relevant binding agreements or soft law instruments. However, here too soft law might be best, as ongoing border issues make reaching a binding agreement unlikely. A broad-scope and multilateral mechanism, like the LMC, has even been suggested as a potential pathway to greater cooperation in the region (Hu, 2016). Regardless of the path forward, binding or non-binding, global or regional, the issue of marine plastics is certain to haunt us for years to come. Given China's increasing importance internationally and regionally, as well as its growing engagement on the global stage, it certainly can play a leading role in tackling this wicked problem.

Acknowledgements

The author would like to thank Professor Owen McIntyre and the editors of this collection for their insightful comments on an earlier draft of this chapter.

References

Agence France-Presse (2019, 23 April). How China's Ban on Plastic Waste Imports Became an "Earthquake" that Threw Recycling Efforts into Turmoil. *South China Morning Post*. Retrieved from https://www.scmp.com/news/china/politics/article/3007280/how-chinas-ban-plastic-waste-imports-became-earthquake-threw

Barresi, P.A. 2017. The Role of Law and the Rule of Law in China's Quest to Build an Ecological Civilization. *Chinese Journal of Environmental Law* 1(1): 9–36.

Boyle, A.E. 1999. Some Reflections on the Relationship of Treaties and Soft Law. *International and Comparative Law Quarterly* 48(4): 901–913.

China-India. 1993. Agreement on Environmental Cooperation between the Government of the Republic of India and the Government of the People's Republic of China (signed and entered into force 7 September 1993).

China-India. 2006. Joint Declaration by the Republic of India and the People's Republic of China (signed 21 November 2006).

China-Mongolia. 1994. Agreement on the Protection and Utilization of Transboundary Waters (signed 29 April 1994).

China-Russia. 2001. Treaty of Good-Neighbourliness and Friendly Cooperation (signed 16 July 2001).

Corne, P. and Zhu, V. 2020. Ecological Civilization and Dispute Resolution in the BRI. *Chinese Journal of Environmental Law* 4(2): 200–216.

Devlaeminck, D.J. 2021. Timeline of the Lancang-Mekong Cooperation (LMC) Mechanism. *Academia. edu*. Retrieved from https://chongqing.academia.edu/DavidDevlaeminck

Devlaeminck, D.J. 2022. Softness in the Law of International Watercourses: The (E)merging Normativities of China's Lancang-Mekong Cooperation. *Transnational Environmental Law* 11(2): 357–380.

Dupuy, P.M. 1990. Soft Law and the International Law of the Environment. *Michigan Journal of International Law* 12(2): 420–435.

Emmerik, T.V. and Schwarz, A. 2019. Plastic Debris in Rivers. *WIREs Water* 7(1): 1–24.

Finska, L. and Howden, J.G. 2018. Troubled Waters—Where is the Bridge? Confronting Marine Plastic Pollution from International Watercourses. *Review of European Community and International Environmental Law* 27(3): 245–253.

Gall, S.C. and Thompson, R.C. 2015. The Impact of Debris on Marine Life. *Marine Pollution Bulletin* 92(1–2): 170–179.

Global Trade (2021, 5 June). China's Recent Ban on Solid Waste Imports to Shift Global Recovered Paper Market, Global Trade Magazine. Retried from https://www.globaltrademag.com/chinas-recent-ban-on-solid-waste-imports-to-shift-global-recovered-paper-market/.

GTI (Greater Tumen Initiative). 1995a. Agreement on the Establishment of the Consultative Commission for the Development of the Tumen River economic Development Area and Northeast Asia (signed 30 May 1995).

GTI 1995b. Memorandum of Understanding on Environmental Principles Governing the Tumen River Economic Development Area and Northeast Asia (signed 30 May 1995).

GTI 2021. List of Documents. Retrieved from http://www.tumenprogramme.org/?list-1527.html

Guzman, A.T. and Meyer, T.L. 2010. International Soft Law. *Journal of Legal Analysis 2*(1): 171–225.

Hu, W. (2016, December 26). LMC can be used as example for China to handle transboundary river disputes. *Global Times*. Retrieved from https://www.globaltimes.cn/content/1025661.shtml

ILC 1988. Report of the Commission to the General Assembly on the Work of its Fortieth Session (A/CN.4/SER.A/1988/Add.1 (Part 2)). *Yearbook of the International Law Commission, II*.

Keselica, T.G. 2020. Fish Don't Litter in Your House: Is International Law the Solution to the Plastic Pollution Problem?. *Pace International Law Review* 33: 115–149.

Kirk, E.A. and Popattanachai, N. 2018. Marine Plastics: Fragmentation, Effectiveness and Legitimacy in International Lawmaking. *Review of European Community and International Environmental Law* 27(3): 222–233.

Kirk, E. 2022. Four Reasons to be Hopeful About the Proposed Global Plastics Treaty. *The Conversation*. Retrieved from: https://theconversation.com/four-reasons-to-be-hopeful-about-the-planned-global-plastics-treaty-178444

Koivurova, T. 2014. *Introduction to International Environmental Law*. Abingdon/New York: Routledge.

Lee, S., Hong, I. and Chung, J. 2019. Transboundary Water Cooperation in the Tumen River Basin: A Focus on the Greater Tumen Initiative. *Asia-Pacific Research* (아태연구) 26(4): 22–54.

LMC 2016. Sanya Declaration of the First Lancang-Mekong Cooperation (LMC) Leaders' Meeting (signed 23 March 2016).

LMC 2018. Phnom Penh Declaration of the Second Lancang-Mekong Cooperation (LMC) Leaders' Meeting (signed 10 January 2018).

LMC 2020. Vientiane Declaration of the Third Mekong-Lancang Cooperation (MLC) Leader's Meeting (signed 24 August 2020).

LMC-MRC 2019. Memorandum of Understanding Between the Mekong River Commission Secretariat and the Lancang-Mekong Water Resources Cooperation Center (signed 17 December 2019).

LOSC 1982a. United Nations Convention on the Law of the Sea (signed 10 December 1982; entered into force 16 November 1994).

LOSC 1982b. United Nations Convention on the Law of the Sea: Status,*UN Treaty Collection*. Retrieved from https://treaties.un.org/Pages/ViewDetailsIII.aspx?src=TREATY&mtdsg_no=XXI-6&chapter=21&Temp=mtdsg3&clang=_en

McCaffrey, S. 2019. *The Law of International Watercourses* (3 ed.). Oxford: Oxford University Press.

McIntyre, O. 2020. Addressing Marine Plastic Pollution as a 'Wicked' Problem of Transnational Environmental Governance. *Environmental Liability* 25(6): 282–295.

Mekong Agreement. 1995. Agreement on the Cooperation for the Sustainable Development of the Mekong River Basin (adopted 5 April 1995).

MFA (Ministry of Foreign Affairs – China) (2016, March 17). Five Features of Lancang-Mekong River Cooperation. Retrieved from www.fmprc.gov.cn/mfa_eng/zxxx_662805/t1349239.shtml

MRC 2020. Actions to address Mekong plastic pollution take shape. Retrieved from http://www.mrcmekong.org/news-and-events/news/actions-to-address-mekong-plastic-pollution-take-shape/

Nguyen, S. (2020, 12 December). Southeast Asia Braces for Trash Dump as China Enacts Waste Import Ban. *South China Morning Post*. Retrieved from https://www.scmp.com/week-asia/health-environment/article/3113665/southeast-asia-prepares-get-dumped-china-enacts-waste

Nikiema, J., Mateo-Sagasta, J., Asiedu, Z., Saad, D. and Lamizana, B. 2020. *Water Pollution by Plastics and Microplastics: A Review of Technical Solutions from Source to Sea*. Retrieved from https://cn.bing.com/search?q=Water+Pollution+by+Plastics+and+Microplastics%3A+A+Review+of+Technical+Solutions+from+Source+to+Sea&cvid=d8e044c0e3314158b2d81f8002d3b51b&pglt=163&FORM=ANSPA1&PC=U531

Nollkaemper, A. 1993. Legal Protection of the Marine Environmnent from Pollution: Recent Developments. *Maritime Pollution Bulletin* 26(6): 298–301.

Oral, N. 2021. From the Plastics Revolution to the Marine Plastics Crisis: A Patchwork of International Law. pp. 281–315. *In*: R. Barnes and R. Long (eds.). *Frontiers in International Environmental Law: Oceans and Climate Change*. Brill.

Parker, L. (2020, July 23). Plastic Trash Flowing in the Seas will Nearly Triple by 2040 Without Drastic Action. *National Geographic*. Retrieved from https://www.nationalgeographic.com/science/2020/07/plastic-trash-in-seas-will-nearly-triple-by-2040-if-nothing-done/

Schmidt, C., Krauth, T. and Wagner, S. 2017. Export of Plastic Debris by Rivers into the Sea. *Environmental Science and Technology* 27(51): 12246–12253.

Semuels, A. (2019, March 6). Is this the End of Recycling?. *The Atlantic*. Retrieved from https://www.theatlantic.com/technology/archive/2019/03/china-has-stopped-accepting-our-trash/584131/

Shelton, D. 2009. Soft Law. pp. 68–80. *In*: D. Armstrong (ed.). *Routledge Handbook of International Law*. New York/Abingdon: Routledge.

Shen, W. and Xie, L. 2019. Can China Lead in Multilateral Environmental Negotiations? Internal Politics, Self-Depiction and China's Contribution in Climate Change Regime and Mekong Governance. *Eurasian Geography and Economics* 59(5–6): 708–732.

Spijkers, O. and Devlaeminck, D.J. 2021. Layers of Regulation in Transboundary Water Governance: Exploring the Role of Third States in the Lancang-Mekong. *Water International* 47(1): 132–151.

Tanzi, A. 2015. *The Economic Commission for Europe Water Convention and the United Nations Watercourses Convention: An Analysis of their Harmonized Contribution to International Water Law (ECE/MP.WAT/42)*. Retrieved from https://www.unece.org/fileadmin/DAM/env/water/publications/WAT_Comparing_two_UN_Conventions/ece_mp.wat_42_eng_web.pdf

Thürer, D. 2009. Soft Law. *In*: *Max Planck Encyclopedia of International Law*.

UNECE. 2013. Guide to Implementing the Water Convention (ECE/MP.WAT/39). Retrieved from https://unece.org/fileadmin/DAM/env/water/publications/WAT_Guide_to_implementing_Convention/ECE_MP.WAT_39_Guide_to_implementing_water_convention_small_size_ENG.pdf

UNEP. 1995. Global Programme of Action for the Protection of the Marine Environment from Land-Based Activities (UNEP(OCA)/LBA/G.2/7) (signed 5 December 1995).

UNEP. 2020. Our Planet is Drowning in Plastic Pollution. *United Nations Environment Programme*. Retrieved from https://www.unenvironment.org/interactive/beat-plastic-pollution/

UNEP. 2021. Understanding the State of the Ocean: A Global Manual on Measuring SDG 14.1.1, SDG 14.2.1 and SDG 14.5.1. https://wedocs.unep.org/handle/20.500.11822/35086

UNEP. 2022. United Nations Environment Assembly of the UNEP, Fifth Session, 2 March 2022 - End Plastic Pollution: Towards an internationally legally binding instrument (UNEP/EA.5/L.23/Rev.1).

UNGA. 1970. General Assembly Resolution 2669 (XXV), 8 December—Progressive development and codification of the rules of international law relating to international watercourses.

UNGA. 1997. 99th Plenary Meeting, Fifty-First Session, United Nations General Assembly (A/51/PV.99).

UNGA. 2016. Transforming our World: The 2030 Agenda for Sustainable Development (A/RES/70/1). https://sustainabledevelopment.un.org/post2015/transformingourworld

VCLT. 1969. Vienna Convention on the Law of Treaties(adopted 23 May 1969; entered into force 27 January 1980).

Vinogradov, S. and Wouters, P. 2013. *Sino-Russian Transboundary Waters: A Legal Perspective on Cooperation*. Retrieved from http://isdp.eu/content/uploads/images/stories/isdp-main-pdf/2013-vinogradov-wouters-sino-russian-transboundary-waters-legal-perspective.pdf

Water Convention 1992a. Convention on the Protection and Use of Transboundary Watercourses and International Lakes (adopted 17 March 1992; entered into force 6 October 1996).

Water Convention 1992b. Convention on the Protection and Use of Transboundary Watercourses and International Lakes: Status. *United Nations Treaty Collection*. Retrieved from https://treaties.un.org/pages/ViewDetails.aspx?src=TREATY&mtdsg_no=XXVII-5&chapter=27&clang=_en

Watercourses Convention 1997a. United Nations Convention on the Law of Non-Navigational Uses of International Watercourses (adopted 21 May 1997; entry into force 17 August 2014).

Watercourses Convention 1997b. Convention on the Law of the Non-Navigational Uses of International Watercourses: Status. *UN Treaty Collection*. Retrieved from https://treaties.un.org/Pages/ViewDetails.aspx?src=TREATY&mtdsg_no=XXVII-12&chapter=27&clang=_en#1

Wouters, P. 2013. *International Law—Facilitating Transboundary Water Cooperation*. Retrieved from http://www.gwp.org/globalassets/global/toolbox/publications/background-papers/17-international-law---facilitating-transboundary-water-cooperation-2013-english.pdf

Wouters, P. and Chen, H. 2013. China's 'Soft-Path' to Transboundary Water Cooperation Examined in the Light of Two UN Global Water Conventions–Exploring the 'Chinese Way'. *Journal of Water Law* 22(6): 229–247.

Xinhua (2020, 1 January). China reveals plan to cut plastic use by 2025. *Xinhua*. Retrieved from http://www.xinhuanet.com/english/2020-01/19/c_138718297.htm

Chapter 14

The Plastic Vote

Referendum as a Governance Tool to Combat Plastic Pollution

Nanuli Silagazde and *Savitri Jetoo**

Direct democracy is a vital element of governance in the United States that has been applied to resolve a wide range of policies. Direct democracy, in contrast to indirect or representative democracy, refers to the direct participation of citizens in the democratic decision-making process. In recent decades only, hundreds of environmental issues have been put on for popular vote addressing, such as issues on energy, forests, natural resources, and water. Despite the widespread use of referendums in this domain, there have been only two popular votes on plastic—in the state of California in 2016 and the city of Seattle in 2009. Both referendums were sponsored by the industry to overturn the previously imposed ban on single-use plastics but came out with two contrasting outcomes. This chapter investigates why these referendums occurred in the first place, the narrative used in the campaigns, and other factors that contributed to their adoption or rejection by the public. Most importantly, it examines whether referendums can be viewed as an effective tool of governance in combatting the problem of plastic pollution. All these aspects gain additional weight amid COVID-19 since hard-won plastic bag bans have been suspended in various communities, and the plastics industry has been seizing the moment and lobbying for further changes in the legislation.

Participatory Democracy in Plastic Governance

In general, elements of direct democracy—citizen initiatives and referendums—have become established tools within environmental governance, especially on the local and provincial levels (Buček and Smith, 2000). Various democratic innovations,

Åbo Akademi University, Social Science Research Institute (Samforsk), Turku, Finland.
* Corresponding author: jsavitri@hotmail.com

such as participatory and deliberative fora, have spread in the field of environmental governance in the hopes of improving environmental policy processes and outcomes (Newig et al., 2019). Apart from direct democracy, North Americans have organised an increasing number of civic forums, including mini-publics, co-governance institutions, and popular assemblies, and 15% of these are devoted to answering environmental questions (Karpowitz and Raphael, 2019).

Direct Democracy in the USA

Given its federal system, the US has traditionally offered its citizens multiple ways for involvement, be it through associations or town meetings. At the turn of the twentieth century, the American Populist Party and its allies fought successfully to establish and spread referendum and initiative (Piott, 2004). Since then direct democracy has become a customary element of American political life. There are three major mechanisms of direct democracy that are widely employed: recall, referendum, and ballot initiative. Given the founders' scepticism about direct democracy, the US Constitution neither authorizes national referendums/initiatives nor allows citizens to vote directly on federal issues (Karpowitz and Raphael, 2019). Accordingly, no ballot can be initiated that refers to legislation that has been approved by Congress; similarly, the recall of elected federal officials is not permitted either (Noyes, 2018). Instead, direct democracy is available at the local and state levels. Although the regulations differ across states, most Americans live in a city or state that allows its citizens to enact laws through the ballot initiative (Matsusaka, 2004). However, the most common restriction for ballot initiatives is related to the topics like votes on budget issues or state institutions (Noyes, 2014). Additionally, most states require that a ballot initiative covers only a single subject in order to prevent voter confusion (Gilbert, 2006). There are significant variations in regard to the signature gathering process. A majority of states have a residency requirement that helps ensure that an initiative has substantial grassroots support within the state and is not hijacked by out-of-state special interests. The amount of time for gathering signatures lies usually between 12 to 18 months. As for the number of signatures that is required to place an initiative on the ballot, there are often considerable differences. Colorado, for instance, requires a number of signatures that are equal to 5% of the votes cast for Secretary of State in the last election, whereas Wyoming's requirement for an initiative is a number of signatures equal to 15% of the total votes cast in the last general election for any office (Noyes, 2014). Interestingly, the five states with the least restrictive rules (California, Oregon, Colorado, North Dakota, and Arizona) account for over half of all state-wide initiatives (Donovan, 2014). On one hand, lower signature requirements along with an open-ended circulation period encourage the use of direct democracy and result in a high number of qualifying ballot initiatives. Also, more ballot initiatives may overwhelm voters and decrease their interest and engagement in the whole process (Magleby, 1984). Furthermore, some citizens abstain from voting when information demands are too great (Donovan and Bowler, 2003).

California is an exceptional case in various regards. First, the modern-day trend of initiatives began in California in 1978 with the adoption of Proposition

13 which cut property taxes from 2.5% of market value to just 1%. Within two years, 43 states had adjusted their taxes (Waters, 2003). California's usage of direct democracy remains exceptionally high. Alone in 2016, 627 local ballot votes were approved and 196 were defeated in Californian municipalities (Schiller, 2018). California is the only state where the legislature may not amend or repeal the ballot initiative unless the initiative itself allows for it (CA Const. art. II, § 9, art. XII, § 2). Moreover, regulations regarding recall are most relaxed here, allowing for its activation from the first day of authority in office (Bowler, 2004). The most prominent case of recall and second recall in the US history took place in California in the year 2003 when Governor Gray Davis of the Democratic Party was removed from office and replaced by Arnold Schwarzenegger of the Republican Party.

On Funding Regulations

Referendum campaigns tend to attract various interest groups. Each of these groups has the potential to influence votes, advance personal agendas, and gain visibility for themselves. Moreover, the involvement of various actors often leads to a disparity of financial resources spent on promoting one referendum option over the other since resources mobilized by different organizations are rarely distributed evenly (Taillon, 2018). This disparity in campaign spending may undermine the quality of the debates due to significant differences in the visibility of each camp in the media via posters, radio, television, etc. (Papadopoulos, 1996). In the absence of legal limits on campaign spending, the very pluralistic nature of the referendum can be jeopardised in the aftermath of reduced diversity and circulation of ideas (Stéfanini, 2004).

Referendums are certainly sensitive to money and financial resources. As Kobach noted, "money unquestionably affects the outcome to some degree" (Kobach, 1994). For instance, in systems where referendums are frequently held—Switzerland and certain states in the US—there is greater professionalisation and commercialisation of signature gathering as specialised firms in this domain offer their services for sale (Cronin, 1989; Kobach, 1994; Magleby, 1984). In fact, the signature gathering has become a big business in the United States. Alone in California during the year 2016, more than $ 45 million was spent to collect 11 million signatures at an average price of almost $ 4 per signature (Mai-Duc, 2016). This means that influential groups have better chances of placing their issues to a vote. However, despite these massive expenditures, approximately 26% of all initiatives that were filed in California had made it to the ballot out of which only 8% were adopted by the voters (Waters, 2003).

In fact, there are different views regarding the role of money in referendum campaigns as the expenditure of funds does not necessarily guarantee victory for one side or the other (Gerber, 1999). Interestingly, differences in spending in the US are more strongly felt in favour of 'No' camp; in other words, it is easier to secure the status quo with substantial resources. In contrast, in Europe, referendums directed towards change are more likely to be adopted compared to those aiming at preserving the status quo (Silagadze, 2022). This means that the financially advantaged camp has a higher chance of beating the proposal rather than succeeding in getting a new measure adopted (Cronin, 1989). Although money plays an important role in shaping the political agenda, it is less powerful at determining the outcome/deciding the vote as measures backed by big corporations to promote their narrow interests more often get rejected than the other kinds of initiatives (Damore et al., 2012; Donovan, 2014).

Despite growing concern over the influence of campaign funding and the importance of its regulation (Gilland-Lutz and Hug, 2010), only a small number of countries have adopted specific regulations. Since referendum campaign legislations are particularly controversial, in many countries such laws emanate from court rulings. Yet, regulations vary substantially across countries. Some countries limit government financing of their side in a referendum campaign, while others only have political neutrality requirements for public authorities. Ireland, for instance, forbids the expenditure of public funds in support of one side of an issue, and France, on the other hand, puts a ceiling on donations (Reidy and Suiter, 2015). In contrast, in Switzerland and the US, despite the potential dangers of deregulated spending, any form of restriction or ceiling on referendum expenditures is viewed through the lens of freedom of expression and is considered an attack on individuals' right to defend their beliefs. The US Supreme Court has ruled that establishing a ceiling on campaign spending would be unconstitutional since controlling campaign spending would inevitably result in the violation of the constitutional principle of freedom of expression. [Similarly, the Constitutional Court in Italy decided that limitations on spending for legislative elections were not applicable to referendum campaigns (Taillon, 2018)].

The Case of Seattle

In the summer of 2008, the Seattle City Council voted to impose a 20-cent fee on paper and plastic bags, which would have taken effect on January 1, 2009. However, the 'Coalition to Stop the Seattle Bag Tax', funded by the American Chemistry Council, the Washington Food Industry, and 7-Eleven, quickly collected 22,000 signatures (instead of the required 20,000) to put the measure on the ballot as Referendum 1 (BALLOTPEDIA, 2009; Thompson, 2009).

According to the adopted legislation, a 20-cent fee would be required for every disposable bag (paper or plastic) distributed by grocery, drug, and convenience stores. Small businesses under $1 million in annual revenue would keep the entire 20-cent fee, while bigger businesses would get to retain 5 cents. The remaining 15 cents would go to Seattle Public Utilities to fund the implementation and oversight of the program, provide free reusable bags to low-income families, and sponsor soup kitchens and homeless shelters (Rucker et al., 2008). The city expected to collect $10 million in annual revenue (Yarow, 2009).The proposed green fee would not apply to bags used in stores to hold bulk items, such as bakery goods, vegetables, nuts, etc., as well as bags used to hold newspapers would be exempted too (Rucker et al., 2008). Seattle was not the first city that attempted to discourage the use of plastic bags, but it was the biggest city in the U.S. that tried to put this kind of surcharge on bags (Kaste, 2009).

The fee was introduced as a way to influence Seattleites' shopping habits, thus encouraging the use of reusable bags without banning disposable ones straight away. Supporters argued that the fee would encourage the use of more reusable bags, cut down on pollution and waste, and reduce greenhouse gas emissions (Yarow, 2009).

The advocates pointed to Ireland's successful PlasTax, which too imposed a similar fee, that resulted in over 90% reduction in plastic bag use only a few weeks after it took effect. The Seattle Public Utilities estimates that each year approximately 360 million disposable bags are used in the city and that most of these are made of plastic. Since plastic bags can take up to 1,000 years to decompose, while paper bags, although easily recyclable, require far more energy and toxic chemicals for their production; one study estimated that Referendum 1 could cut greenhouse-gas emissions by about 4,000 tons per year which was the equivalent of taking 665 cars off the road (Rucker et al., 2008). However, the 'Yes' camp managed to raise only around $64,000—a tiny fraction of the $1.4 Mio raised by the 'No' side. The main organization supporting the bag fee was the Seattle Green Bag Campaign, backed by Mayor Greg Nickels, five Seattle City Council members, National Wildlife Federation, and UW Sierra Student Coalition. Additionally, the Seattle Times endorsed the measure as well as the former president and CEO of Starbucks Orin Smith. The 'Yes' side framed their campaign as a fight against big oil and environmental pollution (C. Thompson, 2009).

The 'No'side was mainly driven by the 'Coalition to Stop the Seattle Bag Tax' and most of its funding came from the American Chemistry Council, whose members included Dow Chemical, ExxonMobil, and major plastic-bag producers. At the same time, some other powerful actors came out against the measure,such as the Independent Business Association, Korean American Grocers Association, Washington Association of Neighborhood Stores, as well as mayoral candidate Jan Drago and Seattle City Council candidates (C. Thompson, 2009). The plastics industry outspent its opponents in their aggressive campaign against the fee with a ratio of about 15 to 1 (Yarow, 2009).

The 'No' side opposed the measure stating that it was costly and unnecessary in a city where plastic bags represented only 1% of the city's garbage and claimed 90% of its citizens were already reusing their disposable bags (Kaste, 2009). Moreover, opponents criticised the legislation included too many exemptions and loopholes to be effective. According to the 'No' camp, a bag tax would not help the environment but could hurt the economy and businesses and even cost jobs. They referred to a measure as a tax, not a fee, although the fact that revenues were designed to go to bag-elimination efforts (instead of into a general fund) puts it more in the fee category (C. Thompson, 2009). One of the arguments of the 'No' camp was that the tax would unfairly affect low-income people and even defended it with the slogan "A tax on grocery bags is not what we need in this economy"(Kaste, 2009).Opponents of the legislation estimated that it would cost Seattle residents $300/year as around 600 bags per person were thrown away/utilised each year. Simultaneously, a nationwide $.20 bag fee in the US would generate $2 billion each year (Fletcher and Hayes, 2009). Additionally, the 'No' camp emphasised the ineffectiveness of the Irish PlasTax, which eliminated plastic shopping bags but did not stop consumers from using plastic bags. In fact, it resulted in a 21% increase in the total amount of plastic consumed in Ireland and a 400% increase alone in sales of garbage bags (PIFA Points To Bag Tax, 2006).

On August 18, 2009, the measure was defeated as 53% of Seattleites voted against it. However, two years later, the Seattle City Council returned with a new proposal and unanimously approved a ban on plastic bags and a 5-cent fee for paper bags, which went into effect on July 1, 2012 (L. Thompson, 2011; Yardley, 2011).

The Case of California

In 2014, California passed the first state-wide legislation in the country banning single-use plastic bags. The legislation was backed by the California Grocers Association and the United Food and Commercial Workers Union. According to the legislation (Senate Bill 270), even recycled paper bags were mandated to be sold for no less than $0.10. It allowed for the continued use of thicker plastic bags that last for 125 uses. The bill, signed by Governor Jerry Brown, was supposed to take effect in July 2015. Larger grocery chains, including Wal-Mart and Target, were required to comply with the law as it entered into force, while smaller convenience, liquor stores, and food marts had an additional year to adjust to the new regulations, which was until July 2016 (Mernit, 2016a, 2016b). Additionally, state plastic bag manufacturers were to be provided with $2 million in loans to help them retain jobs and transition to making thicker and multi-use recycled plastic bags (J. Gerber, 2016).

Taking into consideration that plastic manufacturers earn between $100 and $150 earned million every year alone in California, they were against this measure, to put it mildly (Mernit, 2016b). The national plastics lobby, united under The American Progressive Bag Alliance (APBA), had fought hard to prevent this state-wide ban from being adopted but in vain. However, in the end, the Alliance succeeded in suspending the measure as they gathered enough signatures to put the bill to a referendum as Proposition 67 (Mernit, 2016a).

According to the California Constitution (Article II), a referendum can be called to approve or disapprove a recently (within 90 days) enacted piece of legislation. The minimum amount of signatures required is 5% of the total number of voters in the previous gubernatorial election, which in this case was around 505,000 (Mahoney and Seaward, 2016). The proponents collected over 800,000 signatures out of which 555,000 were verified and thus qualified for the referendum (White, 2014). The costs for signature collections were estimated at around $3 Mio or $5.77 per signature (BALLOTPEDIA, 2016).

In addition, the plastics manufacturers managed to add another initiative to the ballot—Proposition 65—which aimed at redirecting fees grocers would collect for disposable carry out bags to a state wildlife fund (Mernit, 2016a). It was meant to be confusing as the plastic industry intentionally applied the strategy of having two competing proposals on the ballot. The bet was on voter fatigue and the expectation that people would just vote no to both. Indeed, as a survey conducted by *Capitol Weekly illustrated* about 70% of respondents had misperceptions about Proposition 65, believing that when supporting it they were voting against the interests of the plastic industry (Mitchell and Yan, 2016). The representative of the plastic trade association did not even argue that helping wildlife was the point, but rather that the goal of Proposition 65 was merely to highlight where the fee goes after grocers collect it (Mernit, 2016b). Furthermore, the wording of Proposition 67 itself was confusing. The 'Yes' vote would uphold the ban and 'No' vote would overturn it. It is not intuitive that one needs to vote 'Yes' for the law to go into effect. One might even think 'I don't want to veto the law' and mistakenly vote 'No' in that case (Mernit, 2016b). However, it must be acknowledged that California ballots often suffer from problems of clarity as it is not seldom that voters are invited to cast their preferences on dozens of different and highly complex matters (LeDuc, 2015).

The campaign was, therefore, truly 'a battle of uneven'. The camp for repealing the ban on plastic bags had only four contributors, whereas the campaign to uphold the ban was backed by more than 500 organizations, including environmental groups, businesses, community organizations, dozens of cities and counties, and more than 40 newspapers endorsing it. However, in financial terms, the campaign was heavily skewed in the other direction. Plastic bag manufacturers spent over $6 Mio against the ban, whereas the campaign in favour had a budget of less than $2 Mio, out of which over 70% came from the environmental community (Davis, 2016).

It is worth mentioning that bans on single-use plastic bags were not a novelty in the state of California, where San Francisco became the first city to enact the banal ready in 2007. Moreover, since June 2016, over 150 cities and counties in California had their own local carry out bag laws (Farsi & Hansen, 2016). For instance, in Los Angeles, the largest city in California and second largest in the US, single-use plastic bags have been banned since 2014 and a fee of $0.10 is charged for a paper carry out bag ("BALLOTPEDIA", 2016). Thus, at the time of the referendum, more than 40% of California communities were already living without plastic shopping bags through local ordinances, and they did not want to roll back (Davis, 2016).

The 'No' camp was notably using the word 'tax' rather than charge or fee in its claims. Their official slogan was: "Don't be fooled: Not one penny of the bag ban tax goes to the environment". However, using the term 'tax' was simply wrong and misleading since revenue would go to stores and not to a government agency. Moreover, the claim completely ignored the indirect benefit the fee could have on the environment by substantially reducing the amount of plastic bags in circulation and encouraging consumers to shift towards reusable bags/options (Nichols, 2016).

The spokesman for the APBA, Jon Berrier, opposed the ban saying that: "It bans a 100-percent recyclable product produced in America with American labor, that according to the U.S. Environmental Protection Agency accounts for only 0.3 percent of the waste stream" (Mernit, 2016b). Although Berrier was right that plastic bags account for only a small percentage of waste, his statement disregards the fact that plastic is recyclable only in special facilities and otherwise does not decompose in hundreds (sometimes thousands) of years (Wang et al., 2016); hence, putting the oceans and wildlife at risk. On the website of APBA, it said that plastic bags are "the smartest, most environmentally-friendly choice at the checkout counter" (Loftus-Farren, 2016).

In the end, on November 8, 2016, California said no to the plastics lobby's wish list, approving Proposition 67 with 52% and rejecting Proposition 65 with 51%. This was exactly the outcome environmental groups, grocers, and unions had advocated for. However, the millions invested in the campaign and lobbying the legislature during previous years did not go waste since the industry made $15 million in profit alone, while the bag ban was on hold (Mernit, 2016a).

Conclusion: Post-Referendum Developments

After California successfully withstood industry-backed referendum in 2016, its annual consumption of plastic bags was reduced by around 15 billion bags. This is a very positive development, especially when keeping in mind that this state had gone to great lengths to increase plastic bag recycling but was never able to go above 3% of total consumption. Along with approximately 150 municipal bans put in place across California since 2007 and with the help of the state-wide ban, the overall amount of plastic bag litter along the California coastline decreased by around three-quarters in the period from 2010 to 2017 (Dauvergne, 2018). Moreover, a recent study has found that California's ban on plastic bags led to a huge decrease in plastic bag consumption, i.e., 71.5% (Taylor, 2019). In fact, the state went even further in its efforts to combat plastic in 2018 when the state legislature passed bills that mandated the development of microplastics management strategies for both drinking water and California's coastal ocean (Martindale et al., 2020).

Similar effects were observed in other locations. A report comparing periods before and after the ban in San Jose found an 89% reduction in bag litter in the city's storm drain system and about a 60% reduction in the city's rivers, streets, and neighbourhoods (Nichols, 2016). After Washington introduced its 5-cent tax in 2009, bag use dropped there sharply too (Rosenthal, 2013). Plastic bans and fees work, notwithstanding the legislative path taken. According to the report issued by Seattle Public Utilities (2016) on the plastic bag ban ordinance from 2011, the measure has proven effective in reducing the number of plastic bags distributed throughout the city. More specifically, the amount of plastic bags in residential garbage declined by almost 50% and by almost 80% in the commercial waste stream. However, a side effect—increased shoplifting—occurred in Seattle (R. Thompson, 2013). Asimilar phenomenon also followed bans in Hawaii, California, and the United Kingdom. This may be explained by the fact that when plastic bags are no longer common, it is more difficult to determine who paid for their goods (Stephenson, 2018).

During the COVID-19 pandemic,plastic-related legislation has been suspended, resulting in a temporary return to single-use plastic. Retailers in California, for instance, were allowed to hand out free single-use plastic bags for 60 days starting in April. The supporters of this measure—California Retailers Association and the California Grocers Association—argued that reusable bags put supermarket employees at risk of being infected. Other cities went even further; San Francisco, for example, temporarily banned reusable bags altogether (Murphy, 2020). In California, the state-wide ban together with the requirement for plastic bags to contain 40% recycled materials was back in force by late June (Staub, 2020). Similarly, in Seattle, the adopted legislation from 2011 was suspended during the outbreak of the virus. However, according to the Seattle Government (2020), the measures have been reinforced starting from January 1, 2021, i.e., a ban on single-use plastic carryout bags and a charge of a minimum of 5 cents for paper bags.

References

BALLOTPEDIA. 2009. Retrieved November 29, 2020, from https://ballotpedia.org/Seattle_Plastic_Bag_Tax,_Referendum_1,_2009

BALLOTPEDIA. 2016. Retrieved November 11, 2020, from https://ballotpedia.org/California_Proposition_67,_Plastic_Bag_Ban_Veto_Referendum_(2016)

Bowler, S. (2004). Recall and representation: Arnold schwarzenegger meets edmund burke. *Representation* 40(3): 200–212.

Buček, J. and Smith, B. 2000. New Approaches to Local Democracy: Direct Democracy, Participation and the 'Third Sector.' *Environment and Planning C: Government and Policy* 18(1): 3–16.

Cronin, T.E. 1989. *Direct Democracy: The Politics of Initiative, Referendum, and Recall.* Cambridge: Harvard University Press.

Damore, D.F., Bowler, S. and Nicholson, S.P. 2012. Agenda Setting by Direct Democracy. *State Politics & Policy Quarterly* 12(4): 367–393.

Dauvergne, P. 2018. Why is the global governance of plastic failing the oceans? *Global Environmental Change* 51: 22–31.

Davis, M. 2016. How California Became America's First State to Ban Plastic Bags. Retrieved September 2, 2020, from https://www.ecowatch.com/california-plastic-bag-ban-2143461966.html

Donovan, T. 2014. Referendums and Initiatives in North America. pp. 122–161. *In*: *Referendums Around the World: The Continued Growth of Direct Democracy*. Basingstoke: Palgrave Macmillan.

Donovan, T. and Bowler, S. 2003. Are Voters Competent Enough to Vote on Complex Issues. *In*: D.M. Waters (ed.). *The initiative and Referendum Almanac: A comprehensive guide to the initiative and referendum process in the United States*. Durham: Carolina Academic Press.

Farsi, M. and Hansen, J. 2016. Proposition 65: Carryout Bags. Charges. *California Initiative Review (CIR), Article* 16: 1–21.

Fletcher, K. and Hayes, D. 2009. Editorial: The petrochemical industry's profits are our litter. Retrieved August 1, 2020, from https://www.seattletimes.com/seattle-news//2009/07/the-petrochemical-industrys-profits-are-our-litter-.html

Gerber, E.R. 1999. *The Populist Paradox : Interest Group Influence and the Promise of Direct Legislation.* Princeton: Princeton University Press.

Gerber, J. 2016. Vote YES on PROP 67 to Uphold the California Bag Ban! Retrieved November 29, 2020, from https://ncrarecycles.org/2016/08/vote-yes-on-prop-67-to-uphold-the-california-bag-ban/

Gilbert, M.D. 2006. Single subject rules and the legislative process. *University of Pittsburgh Law Review*, 67(4): 803–870.

Gilland-Lutz, K. and Hug, S. (eds.). 2010. *Financing Referendum Campaigns*. Basingstoke: Palgrave Macmillan.

Karpowitz, C.F. and Raphael, C. 2019. Democratic innovations in North America. pp. 371–388. *In*: S. Elstub and O. Escobar (eds.). *Handbook of democratic innovation and governance*. Cheltenham & Northampton: Edward Elgar Publishing.

Kaste, M. 2009. Debate Over Plastic Bags Heats Up In Seattle. Retrieved November 29, 2020, from https://www.npr.org/templates/story/story.php?storyId=111662657&t=1606566014505

Kobach, K.W. 1994. Switzerland. pp. 98–153 *In*: D. Butler and A. Ranney (eds.). *Referendums around the World: The Growing Use of Direct Democracy*. Washington, D.C.: American Enterprise Institute for Public Policy Research.

LeDuc, L. 2015. Referendums and deliberative democracy. *Electoral Studies* 38: 139–148.

Loftus-Farren, Z. 2016. Can California Finally Ditch Single-Use Plastic Bags? Retrieved September 2, 2020, from https://www.earthisland.org/journal/index.php/articles/entry/can_california_finally_ditch_single-use_plastic_bags/

Magleby, D.B. 1984. *Direct Legislation: voting on ballot propositions in the United States*. Baltimore & London: Johns Hopkins University Press.

Mahoney, R. and Seaward, S. 2016. Proposition 67: Ban on Single-Use Plastic Bags. *California Initiative Review (CIR), Article* 18: 1–22.

Mai-Duc, C. 2016. Get signatures, make money: How some gatherers are making top dollar in this year's flood of ballot initiatives. Retrieved November 30, 2020, from https://www.latimes.com/politics/la-pol-ca-signature-gatherers-ballot-initiatives-california-20160627-snap-htmlstory.html

Martindale, S., Weisberg, S. and Coffin, S. 2020. Status of Legislation and Regulatory Drivers for Microplastics in California. *Readabout* 54: 17–22.

Matsusaka, J.G. 2004. *For the Many or the Few: The Initiative, Public Policy and American Democracy, Chicago*. Chicago: University of Chicago Press, p. 15.

Mernit, J.L. 2016a. After the Vote: Plastic Bag Battle Over, More Fights on the Horizono Title. Retrieved October 15, 2020, from https://capitalandmain.com/after-the-vote-plastic-bag-battle-over-more-fights-on-the-horizon-1110

Mernit, J.L. 2016b. Plastic Bag Ban Prop 67. Retrieved September 13, 2020, from https://www.laprogressive.com/plastic-bag-ban-prop-67/

Mitchell, P. and Yan, A.N. 2016. CA120: Crunching the poll numbers, big time. Retrieved November 8, 2020, from https://capitolweekly.net/ca120-poll-numbers-crunching-big-time/

Murphy, H. 2020. California Lifts Ban on Plastic Bags Amid Virus Concerns. Retrieved November 9, 2020, from https://www.nytimes.com/2020/04/24/us/california-plastic-bag-ban-coronavirus.html

Newig, J., Challies, E. and Jager, N.W. 2019. Democratic innovation and environmental governance. pp. 324–338. *In*: S. Elstub and O. Escobar (eds.). *Handbook of democratic innovation and governance*. Cheltenham & Northampton: Edward Elgar Publishing.

Nichols, C. 2016. PolitiFact: Does Prop. 67 money go to environment? Retrieved September 2, 2020, from https://capitolweekly.net/politifact-plastic-bag-prop-67/

Noyes, H.S. 2014. *The Law of Direct Democracy*. Durham: Carolina Academic Press.

Noyes, H.S. 2018. Existing regulations and recommended best practices: the example of the USA. pp. 271–285. *In*: L. Morel and M. Qvortrup (eds.). *The Routledge Handbook To Referendums And Direct Democracy*. London and New York: Routledge.

NS Packaging. 2006. PIFA Points To Bag Tax. Retrieved November 29, 2020, from https://www.nspackaging.com/news/pifa-points-to-bag-tax-failure/

Papadopoulos, Y. 1996. Les mécanismes du vote référendaire en Suisse : L'impact de l'offre politique. *Revue Francaise de Sociologie* 37(1): 5–35.

PIFA, Mike Kidwell Associates. 2006. Retrieved November 29, 2020, from https://www.nspackaging.com/news/pifa-points-to-bag-tax-failure/

Piott, S. 2004. *Giving Voters a Voice: The Origin of the Initiative and Referendum in America*. Columbia, MO: University of Missouri Press.

Reidy, T. and Suiter, J. 2015. Do rules matter? Categorizing the regulation of referendum campaigns. *Electoral Studies* 38: 159–169. Retrieved from http://www.sciencedirect.com/science/article/pii/S0261379415000426

Rosenthal, E. 2013. Should America Bag the Plastic Bag? Retrieved September 1, 2020, from https://www.nytimes.com/2013/05/19/sunday-review/should-america-bag-the-plastic-bag.html

Rucker, R.R., Nickerson, P.H. and Haugen, M.P. 2008. *Analysis of the Seattle Bag Tax and Foam Ban Proposal*. Retrieved from http://www.seattlebagtax.org/ruckerreport.pdf

Schiller, T. 2018. Local referendums: A comparative assessment of forms and practice. pp. 60–80. *In*: L. Morel and M. Qvortrup (eds.). *The Routledge Handbook To Referendums And Direct Democracy*. London and New York: Routledge.

Seattle Government: Bag Requirements. (2020). Retrieved December 1, 2020, from https://www.seattle.gov/utilities/protecting-our-environment/sustainability-tips/waste-prevention/at-work/bag-requirements

Seattle Public Utilities: report on plastic bag ban ordinance. 2016. Retrieved from http://mrsc.org/getmedia/d4703f1e-2ab4-4bfd-9f0a-98ca5ca32b44/s42plastic.pdf.aspx

Staub, C. 2020. California reinstates bag ban and PCR requirements. Retrieved December 1, 2020, from https://resource-recycling.com/plastics/2020/06/24/california-reinstates-bag-ban-and-pcr-requirements/

Stéfanini, M.F. 2004. *Le contrôle du référendum par la justice constitutionnelle*. Paris: Presses Universitaires d'Aix-Marseille.

Stephenson, F.E. 2018. Persecuting Plastic Bags. pp. 351–360. *In*: A.J. Hoffer and T. Nesbit (eds.). *For Your Own Good: Taxes, Paternalism, and Fiscal Discrimination in the Twenty-First Century.For Your Own Good: Taxes, Paternalism, and Fiscal Discrimination in the Twenty-First Century.* Arlington, VA: Mercatus Center at George Mason University.

Taillon, P. 2018. The democratic potential of referendums: intrinsic and extrinsic limitations. pp. 169–191. *In*: L. Morel and M. Qvortrup (eds.). *The Routledge Handbook To Referendums And Direct Democracy*. London and New York: Routledge.

Taylor, R.L.C. 2019. Bag leakage: The effect of disposable carryout bag regulations on unregulated bags. *Journal of Environmental Economics and Management* 93: 254–271.

Thompson, C. 2009. Controversy heats up over Seattle's proposed disposable bag fee. Retrieved November 29, 2020, from https://grist.org/article/2009-08-07-bag-fee/

Thompson, L. 2011. Seattle City Council bans plastic shopping bags. Retrieved November 5, 2020, from https://www.seattletimes.com/seattle-news/seattle-city-council-bans-plastic-shopping-bags/

Thompson, R. 2013. Plastic bag ban leads to nationwide increase in shoplifting rates. Retrieved November 4, 2020, from https://dailycaller.com/2013/07/24/plastic-bag-ban-leads-to-nationwide-increase-in-shoplifting-rates/

Wang, J., Tan, Z., Peng, J., Qiu, Q. and Li, M. 2016. The behaviors of microplastics in the marine environment. *Marine Environmental Research.*

Waters, D.M. 2003. *The Initiative and Referendum Almanac: A comprehensive guide to the initiative and referendum process in the United States*. Durham: Carolina Academic Press.

White, J.B. 2014. California plastic bag ban referendum has enough signatures, backers say. Retrieved November 17, 2020, from https://www.sacbee.com/news/politics-government/capitol-alert/article5122236.html

Yardley, W. 2011. Seattle Bans Plastic Bags, and Sets a 5-Cent Charge for Paper. Retrieved September 2, 2020, from https://www.nytimes.com/2011/12/20/us/seattle-bans-plastic-bags-and-sets-a-5-cent-charge-for-paper.html

Yarow, J. 2009. Seattle Rejects Its Plastic Bag Tax. Retrieved November 29, 2020, from https://www.businessinsider.com/seattle-rejects-its-plastic-bag-tax-2009-8?r=US&IR=T

Chapter 15

Plastic Pollution Treaty

Way Forward

Neha Junankar

Considering the current global plastic waste situation,there are two urgent issues to be addressed; one is to reduce the volume of uncontrolled or mismanaged waste streams going into the water bodies (including oceans) and the other is to increase the level of recycling (UNEP, 2021). There is also a need for a legally binding global treaty. This chapter looks at some of the existing treaties—such as the Basel Treaty and its amendment as well as the End Pollution Treaty—its draft form, and way forward.

Basel Treaty: Amendment 2021

The value chain and lifecycle of plastics cross lots of borders. The plastic life cycle includes extraction of raw materials, design and production, packaging and distribution, use and maintenance, and recycling, reuse, recovery, or final disposal. Plastic pellets—the raw material used to make plastic products—are produced in one region and traded in another jurisdiction to manufacture the end product. Along with consumer goods trading, plastic waste generated has been traded across continents in the past few decades.

From the 1990s until 2011,the trans boundary movement of plastic waste has steadily increased. It is estimated that in 2016, almost half the plastic waste recycled (14.1 million tonnes) was not processed in the country but was exported to other locations (Brooks et al., 2018; UNEP, 2021).

China was the major importer of plastic waste till 2018, almost accepting 50% of globally traded plastic waste for decades until it imposed bans on importing plastic waste. The world's manufacturing hub,China, sent ships full of goods to North America and Europe, and on their way back, it brought back plastic waste from North America as well as Europe to be processed in China. Imported waste plastic

Engineering and Public Policy, Wbooth school of Engineering practice and technology, McMaster University, Canada.
Email: junankan@mcmaster.ca

was then recycled to produce new plastic products to be either exported or consumed by its own industries. This was predominantly processed via an informal sector that provided low-technology processing services to exporters (Velis, 2014; Rucevska et al., 2015). High-income countries such as Germany, Japan, the United Kingdom, and the United States were the top plastic waste exporters (UN Comtrade, 2021). The incoming plastic waste was contaminated with mixed grade and consisted of poor quality. The cheap labour available in the country would segregate the recyclables or higher grades from mixed waste and further send them for processing, while the leftover low grades and contaminants were burnt or dumped in the oceans or landfills. To restrict illegal trade and import of poor-quality waste, China banned imports of plastic waste in 2018. Moreover, China's internal plastic waste production is sufficient by now to fulfil its recycling raw material demand.

By 2019, China's global share of plastic waste imports had fallen by 46%, but new locations for overseas recycling and other disposal operations were opening up. India, Malaysia, the Republic of Korea, Thailand, Indonesia, the Philippines, and Vietnam emerged as contenders (UNCTAD, 2020; UN Comtrade, 2021). There was a sharp increase in illegal waste shipments entering Southeast Asia (Interpol, 2020). The plastic waste is usually misdeclared and traded into the country, where low-grade plastics are disposed of in landfills, rivers, and even plantations and only high-grade would be recycled. Some governments like that of Malaysia took swift action by reinforcing border controls and banning plastic waste imports.

To tackle the illegal trade of plastic waste the Basel convention was amended in 2021. These amendments clarify which types of plastic waste are considered hazardous and which ones, although not hazardous, still require special consideration and are subjected to the control procedure for exports, transit, and import of wastes under the Basel Convention (UNEP, 2021). The Basel Convention Plastic Waste Amendments imply that all plastic waste and mixtures of plastic waste generated by parties to the convention are to be moved to another party, which is subjected to the prior and informed consent procedure unless they are non-hazardous and destined for recycling in an environmentally sound manner and almost free from contamination and other types of waste. Exports to Africa, the Middle East, and South America have increased in the past decade but are still relatively small (Wang et al., 2020). Over time these regions are expected to become more important. The lack of legally binding or lenient enforcement laws in developing countries provides businesses with opportunities to illegally trade waste. In developed countries, the recycling costs are higher along with stricter waste disposal regulations than in developing countries that boosts this trading for recycling.

Many nations have policies and regulations in place to deal with plastic waste. The legislation explorer developed by the UNEP secretariat is a compilation of all these legislations covering plastic pollution. The explorer has 599 National legislations along with subsidiary legislation and policies from around the globe. The Basel Convention is the only global legally binding instrument that controls the trans boundary movement of hazardous waste and its safe and proper disposal with the objective to protect human health and the environment from its adverse effect. In 2019, the Conference of Parties (COP) added plastic waste and amended the convention.

Other Related Global Treaties

The Stockholm Convention regulates a number of Persistent Organic Pollutants (POPs) used as plastic additives. Other chemicals not covered under the Stockholm Convention are considered under a voluntary initiative, the Strategic Approach to International Chemicals Management (SAICM). A recent report (Karasik et al., 2020) lists 28 global policies that have emerged in the last 20 years that address marine pollution. However, few of these policies specifically address plastic pollution or have targets or legally binding commitments to address that issue (especially in relation to land-based sources of marine litter). Thousands of businesses have entered voluntary agreements,such as the Ellen MacArthur Foundation in collaboration with UNEP's Global Commitment and the Ellen MacArthur Foundation's Plastics Pact Network. However, the new 2020 plastic economy global commitment progress report pointed out that current efforts are not sufficient to stop the growing use and pollution problem. Without any major interventions with current demand and under business-as-usual scenarios,production is estimated to ramp up to 1,100 million annually by 2050 (Geyer, 2020), producing 155–265 million tonnes of mismanaged plastic waste.

Need for Global Plastic Treaty

The nature of existing legal instruments across countries is diverse and mostly focused on end-of-life and downstream activities. There are different standards across jurisdictions for designing plastics, so they can be recycled, or have different recycling, and design standards. While some countries have banned the use of certain grades of plastic, neighbouring countries may have not and this may lead to the production, export, and even propagation of illegal trade; thus, lowering the effectiveness of existing efforts made in some parts of the world that advocate the need for harmonized regulations and collective actions. A global treaty will harmonize existing, diverse, and dispersed initiatives and stimulate new policies. Considering how massive the plastic problem is and that it lasts an entire lifecycle, it needs an effective global legal instrument focusing on the entire life cycle. To achieve the systemic transformation, it needs to achieve the transition to a circular economy. The global treaty will enhance investment planning and coordination in infrastructure development as well as technology transfer in developing countries and socio-economically disadvantaged areas where waste management challenges still exist.

Current efforts are primarily committed towards macroplastic (MaP) pollution whereas microplastics or MPs (plastic particles less than 5 millimetres in size) including nanoplastics are overlooked. There is an urgent need for MPs to be considered in regulations and legislation addressing air, water, and soil quality. This will provide impetus to material and process innovation to reduce the number of microfibres from textiles. Multiple independent studies have shown that MPs have made their way into our food chain and the human bloodstream. Some studies estimate that people consume more than 50,000 particles of plastics every year (Cox et al., 2019).

MPs and MaPs may contain hazardous additives. These additives pose a significant threat to our health and ecosystem and are often leaked into marine, air, land, and soil environment during the different stages of the plastic life cycle. A 2018 study found that 7,000 additives are used globally,such as fillers, flame retardants, plasticizers, dyes, etc. (Flick, 2004); many of these are hazardous at different levels to the humans and environment. Out of these, especially have been identified as the most hazardous (Groh et al., 2018). Some additives are regulated under various chemical regulations, such as phthalates. Phthalates are plasticizers that are used in a wide range of applications such as self-care products, toys, electronics, electrical equipment, etc. EU ECHA classified 4 phthalates as toxic to reproduction as these can damage human fertility and affect unborn children,hence its use needs to be regulated. USEPA, under TSCA, also regulates the use of phthalates. These chemicals enter our food chain and in an environment that is often overlooked as the focus is more concentrated on the pollution or downstream activities of plastics' value chain.

End Plastic Pollution: Towards an Internationally Legally Binding Instrument (Current, Draft Resolutions, and Adopted Resolution)

At the United Nations Environmental Assembly (UNEA-5), held on March 02 2022, in Nairobi, Kenya, heads of states, ministers of environment, and other representatives from the UN member states unanimously agreed to have a global plastic pollution treaty—a legally binding agreement to end plastic pollution. There solution 'End Plastic Pollution: Towards an Internationally Legally Binding Instrument' was adopted after a three-day meeting at Nairobi with more than 3,400 in-person and 1,500 online participants from the UN member states. Inger Andersen, executive director of UNEP, said the following:

> Today marks a triumph by planet earth over single-use plastics. This is the most significant environmental multilateral deal since the Paris accord. It is an insurance policy for this generation and future ones, so they may live with plastic and not be doomed by it.

The final a doptedre solution is based on three initial draft resolutions from Rwanda-Peru presented in September (2021) followed by Japan in December (2021) and India in January (2022). These initial resolutions have some commonalities and differences at various levels. A joint frame work was presented by Rwanda and Peru at the Ministerial Conference on Marine Litter and Plastic Pollution in September 2021 to address plastic pollution followed by Japan presenting its resolution in December 2021. Both drafts proposed legally binding actions along with a call for quick negotiations and included the need for technical and financial support mechanisms for developing economies and socio-economically disadvantaged areas with special considerations for the economies in transition. While Rwanda-Peru resolution has the full life cycle approach addressing plastic pollution in all compartments of the environment on land, water, and the air, Japan's draft resolution focused more on marine plastic pollution and down stream interventions. Both drafts suggested setting up an international negotiating committee to prepare the treaty.

The third resolution by India proposed a voluntary approach focusing on single-use plastics. India had already put a resolution ahead in UNEA-4 (2019), which it recalled and released another draft month ahead of UNEA-5. As the draft emphasized concerns about the growing use of single-use plastic (SUP),it elaborated on the role of short-lived plastic packing and its upward demand. India stressed the need for all the relevant stakeholders to get involved and asked member states to implement extended producer responsibility (EPR), make provisions to replace plastic packing with sustainable and greener options, develop policies to encourage recycling, and improve resource efficiency; India also requested member states to prepare national/regional action plans, share best practices, and enhance cooperation in exchanging scientific research, data, and technology on sustainable alternatives to SUP. It shared its regional framework on EPR, including countries' collection, recycling and reuse targets, and notified its decision to ban most single-use plastic by July 2022.

India pushed for "common but differentiated responsibility" as it wants to focus on the waste management/diversion aspect,tackle reducing plastic footprint and fossil fuel for production later, and have the right to decide on the timeline for that. The 60 member states were in favour of the draft submitted by Rwanda and Peru in the first meeting. Norway and the EU were strong supporters of Peru's draft, while Japan, India, and the US opposed it strongly but agreed with the addition of certain items. One point was dropped from negotiations as the US strongly objected, i.e., chemicals of concern in the plastic. There have been no concrete discussions in negotiation on whether there will be a restriction on fossil fuel use (carbon emission obligations during the use phase), but the most carbon-intensive phase is the extraction phase. There is an expectation to see a lot of resistance from the petrochemical industry. The emphasis was also on the developed countries having an advanced collection and sorting system in place but not beyond that as they export which has not solved the crisis and there should be national targets to mitigate by addressing the life cycle than by just managing waste which would be easy for trading.

Few Quotes From Participating Governments After Agreeing to Develop Historic Global Instruments

Modesto Montoya, Minister of Environment Government of Peru said, "We appreciate the support received from the various countries during this negotiation process. Peru will promote a new agreement that prevents and reduces plastic pollution, promotes a circular economy, and addresses the full life cycle of plastics."

A quote from Dr. Jeanne d'Arc Mujawamariya,Rwanda's minister of the environment:

The world has come together to act against plastic pollution—a serious threat to our planet. International partnerships will be crucial in tackling a problem that affects all of us, and the progress made at UNEA reflects this spirit of collaboration. We look forward to working with the INC and are optimistic about the opportunity to create a legally binding treaty as a framework for national ambition-setting, monitoring, investment, and knowledge transfer to end plastic pollution.

Following is a quote from the Government of Japan. "The resolution will take us towards a future with no plastic pollution, including in the marine environment," said Tsuyoshi Yamaguchi, Japan's Environment Minister, whose draft resolution contributed to the final resolution. "United, we can make it happen. Together, let us go forward as we start the negotiations towards a better future with no plastic pollution."

Final Resolution and Way forward

The joint resolution drafted by Rwanda and Peru was favoured but the final resolution includes some elements from Japan and India's resolution as well. The final treaty would have binding and voluntary approaches and actions, focusing on the entire life cycle of the plastics that have a circular approach. The adopted resolution recognises plastic pollution includes MPs. Implicating it may have measures related to MP pollution along with visible MaP pollution. Though the treaty considers pollution at all levels on land, air, and water, special attention will be given to marine litter. Setting up an intergovernmental negotiating committee to come up with the treaty is also necessary.

Few Highlights From the Adopted Draft

"Acknowledges that some legal obligations arising out of a new international legally binding instrument will require capacity building and technical and financial assistance to be effectively implemented by developing countries and countries with economies in transition".

Dedicated multilateral funds:

> To develop, implement and update national action plans reflecting country-driven approaches to contribute to the objectives of the instrument; To promote national action plans to work towards the prevention, reduction, and elimination of plastic pollution and to specify national reporting, as appropriate.

Timeline for Final Treaty

As 175 countries agreed to create a legally binding global treaty at UNEA-5 on March 2, 2022, the next step was to set up an intergovernmental negotiating committee; the executive director of UNEP and the committee then commenced its work during the second half of 2022, which will go on till the end of 2024. INC is expected to complete a draft of a global legally binding agreement.

The resolution also mentions Executive Director of UNEP convene an ad-hoc open-ended working group to hold one meeting during the first half of 2022 to prepare for the work of the intergovernmental negotiating committee in particular to discuss the timetable and organization of work of the intergovernmental negotiating committee. INC will report the progress in UNEA-6. Draft stresses the need to ensure the widest possible and effective participation in the ad-hoc open-ended working group meeting and the intergovernmental negotiating committee; requests the Executive Director as a priority action to provide the necessary support to developing countries and countries with economies in transition to allow for effective participation in the work of the ad-hoc open-ended working group meeting and the intergovernmental negotiating.

What Can be Expected

Experts say there can be chances that industry associations and producers would try to add options, such as biodegradable plastics and plastic-to-fuel, and this will enable producers to claim the plastics can be degraded or used as fuel and label it as circular and get it licensed to ramp up production. This could be difficult and may threaten the objective of decreasing production.

Not sure if we can expect similarities with the carbon economy, like plastic offset, plastic neutrality and plastic washing (like green washing), and trading plastic credits. Or if there will be measures already thought and incorporated from lessons that were learned during the Paris agreement, especially the consequences of carbon trading.

References

Freinkel, S. 2011. A Brief History of Plastic's Conquest of the World Cheap plastic has unleashed a flood of consumer goods. Scientific American. https://www.scientificamerican.com/article/a-briefhistory-of-plastic-world-conquest/.

GESAMP 2015. Sources, fate, and effects of microplastics in the marine environment: a global assessment (Part 1). IMO/FAO/UNESCO-IOC/UNIDO/WMO/IAEA/UN/UNEP/UNDP Joint Group of Experts on the Scientific Aspects of Marine Environmental Protection.

Geyer, R. 2020. Production, Chapter 2- Production, Use, and Fate of Synthetic Polymers in Plastic Waste and Recycling. Letcher, T.M. (ed.). Cambridge, MA: Academic Press, pp. 13–22.

Keswani, A., Oliver, D.M., Gutierrez, T., and Quilliam, R.S. 2016. Microbial hitchhikers on marine plastic debris: Human exposure risks at bathing waters and beach environments. pp. 10–19. *In*: Marine Environmental Research. 118. https://doi.org/10.1016/j.marenvres.2016.04.006

Lamb, J.B., Willis, B.L., Fiorenza, E.A., Couch, C.S., Howard, R., Rader, D.N. et al. 2018. Plastic waste is associated with disease on coral reefs. *Science* 6374, pp. 460–462. https://doi.org/10.1126/science.aar3320.

Maes, T., Barry, J., Stenton, C., Roberts, E., Hicks, R., Bignell, J. et al. 2020a. The world is your oyster: low-dose, long-term microplastic exposure of juvenile oysters. Heliyon 6(1): e03103. https://doi.org/10.1016/j.heliyon.2019.e03103.

Maes, T., van Diemen de Jel, J., Vethaak, A.D., Desender, M., Bendall, V.A., van Velzen, M. et al. 2020b. You Are What You Eat, Microplastics in Porbeagle Sharks From the North-East Atlantic: Method Development and Analysis in Spiral Valve Content and Tissue. *Front. Mar. Sci.* https://doi.org/10.3389/fmars.2020.00273.

Parker, L. 2020. The world's plastic pollution crisis is explained. https://www.nationalgeographic.com/environment/habitats/plasticpollution/.

Rodrigues, A., Oliver, D.M., McCarron, A. and Quilliam, R.S. 2019. Colonization of plastic pellets (nurdles) by *E. coli* at public bathing beaches. Marine Pollution Bulletin. 139, pp. 376–380. https://doi.org/10.1016/j.marpolbul.2019.01.011.

United Nations Environment Programme (UNEP) (2021). *Drowning in Plastics—Marine Litter and Plastic Waste Vital Graphics*. ISBN: 978-92-807-3888-9.

Zettler, E.R., Mincer, T.J. and Amaral-Zettler, L.A. 2013. Life in the "plastisphere": Microbial communities on plastic marine debris. *Environmental Science and Technology* 47(13): 7137–7146. https://doi.org/10.1021/es401288x.

Index

About the Editors

Dr. Gail Krantzberg works at the interface of science and public policy with extensive expertise in disciplines involving remediation of degraded freshwater ecosystems, regeneration policies for natural ecosystem assets, and the advancement of Great Lakes resilience. She is Professor of the Masters of Engineering and Public Policy at McMaster University. She worked for the Ontario Ministry of Environment from 1988 to 2001 and was the director of the International Joint Commision's Great Lakes Regional Office from 2001–2005.

Dr. Savitri Jetoo is an Adjunct Professor at Åbo Akademi University in Finland, whose work was recognised by the Baltic Sea Science Day Award. She obtained her undergraduate degree at the University of Queensland, Australia and her doctorate at McMaster University, Canada. She works at the interface of science and public policy with extensive expertise in water resources governance in many areas of the world where she lived and worked including the Great Lakes Region.

Dr. Velma I. Grover has work experience spanning over two decades in international development with international policy think-tank, non-governmental sector, consulting and teaching at Universities (McMaster University and York University, Canada, and Kobe College, Japan). Her research interests include water, waste and impact of climate change on water cycle and health, and regeneration policies. She teaches at McMaster University in the 'Masters of Engineering and Public Policy' program.

Dr. Sandhya Babel is a full professor at the School of Biochemical Engineering and Technology, Sirindhorn International Institute of Technology, Thammasat University, Thailand. Her main research focus has been on development of low-cost technolgies for the protection of environment. Specific research interests include microplastics pollution and uptake by biota, microbial fuel cells, adsorption, photocatalysis, phytoremediation, and membrane technology. She has been listed in the top 2% scientists in the world by Stanford University and Elsevier BV since 2020.